Salomon Frederik Van Oss

American Railroads and British Investors

Salomon Frederik Van Oss

American Railroads and British Investors

ISBN/EAN: 9783337416768

Printed in Europe, USA, Canada, Australia, Japan

Cover: Foto ©berggeist007 / pixelio.de

More available books at **www.hansebooks.com**

AMERICAN RAILROADS

AND

BRITISH INVESTORS.

BY

S. F. VAN OSS,

AUTHOR OF "AMERICAN RAILROADS AS INVESTMENTS."

LONDON:
EFFINGHAM WILSON & CO., ROYAL EXCHANGE,
1893.

CONTENTS.

CHAP.		PAGE.
I.—The American Railroad of To-day	-	1
II.—American Railroad Finance	- -	26
III.—Capitalisation	- - - - -	36
IV.—Rates and Revenue	- - - -	56
V.—Revenue and its Application	- -	87
VI.—Securities and their Returns	- -	108
VII.—Summing up the Case	- - - -	130
VIII.—How to Invest	- - - - -	154

Appendices.

PREFACE.

THE favourable reception accorded to my recent work on American Railroads has induced its London Publishers to suggest to me that I should write another volume of less formidable dimensions,—an abbreviated popular edition, as it were. I have the more readily accepted this proposal, because the scope of the larger book precluded me from entering as fully as was, perhaps, desirable into various matters which are always of great importance, but which recently have been forced still more into the foreground by one or two sensational events. Thus the present book is not precisely a condensed edition of its more bulky predecessor. It deals with railroad conditions and investments in general, whereas the chief object of my earlier effort was to desscribe the history and position of the various companies in particular; and it discusses at length various matters which received only superficial treatment in my other book.

The principal of the sensational events to which I have just alluded is, of course, the Reading collapse. It is most intimately connected with one of the chief features and one of the greatest defects of American railroad management, namely, the great discretionary powers of the directorates. These powers are of such importance to the investor that I have made them the keynote of the following pages; and although I have introduced no new arguments, I trust that the more or less exhaustive discussion which follows will be found serviceable, and deprive the characteristic dealt with of the greater part of the horror with which it is undoubtedly regarded by the average investor. Its drawbacks are almost entirely counterpoised by the peculiarities of capitalisation which it engendered, and investors need not expose themselves to its influences unless they like. There are hundreds of securities shielded against any possible abuse of discretion on the part of the managers.

There is another matter to which I have been careful to call attention, and that is the great difference between the merits of one security and the other, and the existence of good and bad securities side by side. To enlarge upon such a theme seems somewhat superfluous, for it ought to be plain to everybody that in a vast field of investments like American railroad securities, stocks of the most varied merits can be found. But unfortunately this fact does not seem to be generally recognised, and especially when something goes wrong the most ludicrous conclusions are drawn from one company or one stock to another. A large number of newspapers have, for example, been induced by the recent Reading *débacle* to condemn American railroads in general, and to discourage investment in any Transatlantic railroad stock. But granted that the Reading had the most depraved of managements, what does that prove against, say, New York Central securities? Certainly not more than the default of Greece would prove against British Consols. Yet one company is made to suffer for the sins of another; and it is therefore not altogether superfluous to call attention to the fact that it is far better, as it is certainly more scientific, to find distinctions and differences, than to draw forced comparisons and to offer nonsensical generalisations.

The following pages further contain a rather full discussion of the Rate question, which is constantly passing through new phases. Space has also been found for a short description of railway conditions in general, for a specification and classification of the numerous kinds of securities, and for a few hints to investors. The Appendices contain data which I have little doubt will prove useful to those for whom they are intended.

S. F. VAN OSS.

LONDON, JUNE, 1893.

AMERICAN RAILROADS & BRITISH INVESTORS.

CHAPTER I.

THE AMERICAN RAILROAD OF TO-DAY.*

Although it is our intention to confine ourselves strictly to those matters which are of direct interest to investors, it seems desirable to give, by way of introduction, a brief sketch of the position which the railroads occupy in America.

To the United States railways have always been of greater relative importance than to other countries. The scarcity of ordinary highways, canals and rivers effectually prevented the development of vast and fertile stretches of country; there was no Baltic or Mediterranean cutting deep inland, and giving access to the heart of the continent by means of large rivers carrying craft to and from the centres of production; there was one vast region, over 3,000,000 square miles in extent, the greater part of which would have been inaccessible even if the grand Mississippi system had not led to what, from a commercial point of view, is a very bad part of the coast. The Father of Streams leads to the South, but the centre

---

* This chapter is part of an article which originally appeared in the "Investors' Review" for August, 1892. I am indebted to the editor of that periodical for the leave he has given me to reproduce it here.

of population and of commerce being in the East, while agricultural production, the mainstay of the States, centres in the West, trade, which in the United States has practically but one direction, could not avail itself of the great river system; and before commerce could be established artificial channels had to be created along which it could move.

The lack of highways of trade was keenly felt, more keenly than in any other country; in fact, the full development of the vast territory west of the Alleghany mountains would have been an impossibility, had not this want been provided for. But, however apparent its necessity, and irrespective of the good prospects of such an enterprise, it may on the whole be regarded as doubtful whether in the early days of railways private enterprise would have embarked upon a similar venture had it not been supported by aid from without. Fortunately such assistance was given, chiefly owing to that intense jealousy between States and cities which in America has been, and still is, the direct cause of innumerable improvements and of amazingly rapid progress.

In the early part of the century the Erie Canal, connecting Lake Erie with the Hudson, had been constructed by the State of New York, and its opening in 1825 soon indicated that the struggle for commercial supremacy would be decided in favour of New York. The other cities then striving to become the great commercial centre of the country, did not, however, look on helplessly. As early as 1823 the State of

Maryland and the City of Baltimore planned the completion of a tramway from Chesapeake Bay to the Ohio River, and Pennsylvania made an unsuccessful attempt to construct a canal and portage railway system that would traverse the State from East to West, and cross the ridges of the Alleghanies. When the Liverpool and Manchester Railway had been opened, these inadequate means of transportation rapidly developed into roads with steam traction, still supported by the States in some form or other. The Baltimore and Ohio and the Erie received advances from the Legislatures in Annapolis and Albany; the State of Pennsylvania owned and worked the Central Pennsylvania Railroad, which was afterwards acquired by the present Pennsylvania Railroad Company; the State of Illinois gave the Illinois Central Railroad Company a strip of land twelve miles broad, and traversing the entire State from Cairo to Chicago; and nearly all the earlier routes received support in some form or other.

The success of these railways was so great that construction soon began to assume considerable proportions. In 1835, 1,100 miles had been built; in 1840 there were 2,800; and in 1850, 9,000 miles; and everywhere new lines were projected and their sponsors encouraged by land grants, immunity from taxation, and other favours. The nation began to see that the presence of a railway meant production and accumulation of wealth, and that its absence rendered development impossible. As a result, railway construction was encouraged by every available means, and, favoured and fostered, patronised and protected by the people and their Government, railways were bound to grow at that rapid rate which has astonished the world.

It cannot be said that the railway corporations have shown much gratitude for the favours showered upon them by the Legislatures. In those days there was in the United States a scramble for wealth which had anything but a salutary effect upon commercial morality; and railway corporations especially, which from the outset attracted the best brains of the nation, were conducted by a class of men too shrewd not to take advantage of the high favour wherewith the people regarded their ventures. The liberality of the Federal, State, and Municipal Governments, and the encouragement and confidence of the public, was in many instances abused, and, instead of leading to gratitude, seems to have aroused that greed which led to fraud, extortion, and blackmail. For not only was the construction of railways taken advantage of for the purpose of illicit gain at the cost of the investor —a matter to which we shall return later on—but railways abused the general desire to promote and accelerate their construction by making peremptory demands for loans or gifts of money from States, counties, and towns. These demands were the more repulsive because those who made them usually laid stress upon the "necessity of changing the routes unless such aid be forthcoming," such alteration being possible since no company was ever required to file a detailed plan with any legislative body, as is necessary in almost every country. Strange to say, such great benefits were generally anticipated from the advent of railways that very few local authorities hesitated to respond to similar pressing requests, and that Municipal and State bonds were issued with a readiness which afterwards brought numerous townships into financial straits, and many a county into difficulties.

But the railways by no means stopped short at extorting money from municipalities. It was not long after they had reached their teens that in their intercourse with individuals they gradually adopted unjust, unlawful, and sometimes even criminal practices, which were far more baneful in their consequences than the wholesale blackmail of counties and municipal corporations. By means of discriminations, pools, and corruption, they inaugurated an era of commercial demoralisation which even moderate writers like Professor T. R. Ely, of Yale College, have denounced in the strongest terms, and which probably is without a parallel in the commercial annals of any nation.

Although pools and a widespread corruption were also strongly disapproved of, public indignation was roused chiefly by differential tariffs, and although it is not necessary to enter into many details, a few of the principal features of the objectionable use of "discrimination" may be mentioned.

There were discriminations against localities, individuals, and classes of freight, and the two first-mentioned species were by far the worst. "Local discrimination" may be defined as the practice of making very low rates for competitive points, while intermediate ones would have to pay excessively high charges. To quote the most noted instance, the rate for cotton from Winona to New Orleans, 100 miles, was $3·25 per bale; from Memphis, 325 miles, only $1·00 was charged, and there were myriads of similar cases, some of them excusable on the ground of water or rail competition, others without any apparent reason. Now everybody who is acquainted with railway business knows that differential tariffs are a necessary evil, but it is evident that their application on too extensive a scale must be highly injurious to the community; indeed, the serious decline in the value of land and the desertion of rural districts in the United States have been attributed to differential railway tariffs.* But in spite of their serious consequences, local discriminations were far less condemnable than discriminations between individuals; persons in the same town would have differential rates, and the differences would often amount to so much that the favoured persons were enriched and the others ruined. By a system of "rebates," "drawbacks," and "underbilling," the

---

* Between 1870 and 1880 land values in seven Eastern States, according to the Tenth Census Report, declined 350,000,000 dollars. This fall has been attributed to discriminations in favour of the West.

grossest injustices were committed; for instance, the Standard Oil Company once received $10,000,000 in eighteen months as rebate for petroleum *shipped by others*, and its monopoly and wealth are well known to have originated in nothing but differential rates which ruined all its rivals by directly transferring money from their pockets to the safe of what was then the Southern Improvement Company. Two of the greatest Chicago "packers" are likewise generally conceded to owe their wealth chiefly to differential tariffs; indeed, thousands made vast fortunes by the undue and criminal preference railways gave them, and myriads of others were ruined in consequence. It should not be imagined that these statements are exaggerated. In "The Railroad Problem," Mr. A. B. Stickney has shown, by a simple calculation, that an Omaha wheat merchant who has a Chicago rate of a quarter cent. per bushel less than his rivals, must ultimately monopolise the business; and the Hepburn Committee, appointed by the New York Chamber of Commerce, collected evidence of over 5,000 cases of discrimination between individuals, practised by the New York Central and Erie Railways in New York State alone.

The proportions of this abominable system increased vastly after 1870, and in 1884 there must have been at least 100,000 prominent traders enjoying rates which gave them unjust and undue preference above others, and which doomed thousands to ruin. Nor did discrimination stop at goods traffic. The "free pass" abuse was at one time so universal that nearly half the number of passengers carried rode free, and the inference from this is that the other half practically paid double fares. In addition, the attempts at "pooling," made in the meantime with the object of staying the decline in rates which resulted from excessive competition, imposed a tyranny not less objectionable than discrimination, and perhaps more hurtful. Pools deprived the community of the benefit of free competition, the soul of trade; moreover, they caused severe fluctuations in transportation charges. These were decidedly detrimental to business in general on account of their short duration, which was connected with the speculative designs of their members—designs directly responsible for the frequent rate-cutting characteristic of the era of pools. Further, the inflation of capital, to which reference is made below, and of which it was alleged that it necessitated excessive rates, as well as the dishonesty of managers, who were suspected of resorting to discrimination for the sake of self-enrichment, was universally condemned, and increased the indignation gradually provoked by so many causes for just complaint.

This systematic discrimination could not fail to provoke angry protests in a country devoted to justice and dedicated to the proposition that all men are equal; but it cannot be said that in its earlier stages resistance was very formidable. For a long time the railways remained so powerful that they could consider it unnecessary to pay any heed to their enemies.

Their favours had made friends among the public; their recognised power rendered it dangerous for a person to oppose them; their free pass bribed press and pulpit, bar and bench; their influence and money bought Legislatures, into which they were the first to introduce that pernicious system of jobbery which still distinguishes American politics. With so many odds against them the enemies of the railroads had small chance of success. Indeed, it took the opposition two decades to attain maturity and to outnumber the powers that were. But their number grew constantly and at the same ratio as did the indignation caused by blackmailing, discrimination, free passes, pooling, stock-watering, and corruption of Courts of Justice and Congress-men, and it can surprise nobody that the growing number of opponents, embittered by the hopelessness characterising their cause for so many years, ultimately were as unreasonable in their claims for remedies as the railways had been unjust in their actions. To find the happy medium between the alleged rights of the railways and the angry demands of an indignant public constituted, indeed, that which had gradually become known as "the railroad problem."

Opposition was bound to come into existence first where the largest class suffered most, and hence it originated in the agricultural States of the North-West, where the Grangers, now merged into the Farmer's Alliance, controlled the Legislatures. In 1871 the State of Minnesota passed the first "Granger

Law," and two years later Wisconsin followed suit, both States forbidding discrimination and fixing maximum rates, thereby setting an example which was afterwards followed by several other States. The railways, however, deliberately disregarded these enactments on the ground of their alleged unlawfulness, and in 1874 President Mitchell of the St. Paul, in a now famous letter to the Governor of Wisconsin, gave his reasons for doing so, the principal of which was that the State had no right to interfere with the charges any corporation chose to make for its services. In consequence the case was carried before the Supreme Court, which in 1876 decided in favour of the State. This decision was of the greatest importance, because it defined the legal position of an American railway, and pronounced it to be, not a private corporation, but a public highway subject to the supervision of the Government. If it were a private corporation its charter would be unlawful, because the "right of way" may not be granted to any enterprise which is not for the public weal; and railways being invested with public use the Government has the power to regulate them.

While this principle decided future legislation, and for that reason was of paramount importance, the decision based upon it was of little use. The railways, pretending that if the Supreme Court could compel them to obey Legislatures it could not force them to run trains at a loss, made reprisals in the shape of a very bad and slow service, and thus the Grangers were driven to a compromise; they relaxed their laws and nominated Commissions to fix more reasonable tariffs. But by the time these Commissions were elected, the railways had bribed the Legislatures, in consequence whereof the new laws were not very dangerous. Nevertheless they were adverse to the railroad interest, although less stringent than the old enactments; and in the meantime the struggle had attracted universal attention, and brought forth innumerable books and pamphlets. Moreover, the Supreme Court soon decided that no State could make laws for Inter-State railways; and as in the meantime the agitation had grown constantly, Congress in 1885 appointed a Committee, presided over by Senator Cullom of Illinois, to enquire into the matter. This Committee presented its report in 1886, adding a rider to the effect that in its opinion "no question of Government policy was of so much importance as this." A fierce struggle in Congress was the result of this report, for the railways of America, like those of England, are well represented in the Legislatures; nevertheless the Inter-State Act was passed in 1887, and discrimination and pooling strictly forbidden.

The railway interest anticipated the gravest results from this Act, the most prominent railroad men, among them Mr. Chauncy M. Depew and Mr. H. V. Poor, predicting general disaster; and no doubt these prophecies would have been fulfilled had not the Act, fortunately for the railroads, possessed such weaknesses as prevented it from being effective. The

principal among them was pointed out to me by Mr.
J. D. Springer, a former judge who is now second
Vice-President of the Atchison Railroad Company,
and consists in the fact that every official can refuse
to give evidence on the ground that by doing so he
might incriminate himself, and no American Court
can require a witness to do that.* But of course the
evidence of officials is necessary unless the Act is to
be no more than a dead letter; and since it can be
refused on the ground of its requirement being contrary to the cardinal principles of American jurisprudence, the Act was bound to be a failure, especially
as there were other circumstances which tended to
render it inefficient. For the railways it is fortunate
that was so, because there can be no doubt that the
Act would have been most disastrous if applied rigourously; its prescriptions, fixed as they were by people
hostile to the railway interest and totally ignorant of
railway practice, were much too severe. As it is, with
all its weakness it did some harm which in a measure
reacted upon the nation at large, for it is natural
that anything detrimental to an interest representing
one-tenth of the nation's wealth, and employing a vast
proportion of its capital, must be injurious to a community with which it is closely interwoven. But the
Act has done some good too. The railways have been
taught that it is dangerous to provoke the anger of
the public, and there can be no doubt that abuses and

---

* Since this article was written steps have been taken to remedy this defect.

the complaints engendered by them have become much more rare since the Act came into force. At the same time the public has seen that it went too far, and that a moderate amount of discrimination, of "watering," and even of "pooling," is inseparable from railway business. Both parties have therefore received a wholesome lesson, and the result is that the railways have abolished many of their most nefarious practices, and are gradually purifying their business methods, while the Legislatures are assuming a less hostile attitude and applying their laws less rigourously; indeed, I have been assured by many authorities in the North-West that the relations between the various legislative bodies and the companies promise to become even cordial as time goes on, and many a State formerly ultra-hostile to the companies certainly has already seen the error of its ways, and amended them. Iowa, for instance, was at one time one of the most vehement opponents of the railways; the consequence was that railway construction became stagnant in that State, and that it stood still while its neighbours progressed and prospered; to-day Iowa is less "hard" on the railways than any Western State. Hence it may be confidently expected that the friction, wherever it exists, will soon wear itself out; the animosity is fed chiefly by abuses and tyranny on the part of the railways, and the latter are not likely to resume their bad practices. Yet it may be useful to mention that everything that can be interpreted as a sign of the revival of tyrannical tendencies may rouse and revive indignation. For example, developments like the Reading combination can never have a salutary effect upon the railway question. It need not necessarily be actual tyranny; the appearance of it is sufficient to provoke the indignation of a nation, by instinct and inclination opposed to the tyranny of trusts and monopolies, and by habit hostile to great corporations.

The matters just referred to, however much interest may be attached to them, only fill the background of this sketch; the foreground is occupied by objects of greater direct importance to the investor, namely the relations—firstly, between the corporations mutually, and secondly, between them and the investor.

The *entente* between the several companies has only lately commenced to become cordial, but even to-day it is far more strained in the United States than in any other country. Intensity in every respect is the great characteristic of the States and of everything American, good and bad alike, and this trait frequently leads the nation or the units of which it is composed into excesses. Among the numerous consequences of this characteristic, the love of "booms" fills no subordinate place; and it was only natural that as soon as the profitable nature of railway enterprise became generally recognised, construction went too far. The people were not satisfied with merely meeting the demand for transportation facilities, but they would persist in anticipating it. Hence railways soon were "boomed for all they were worth," and as a result the supply of transportation was speedily considerably in excess of the demand; to-day the disproportion still exists in a more or less marked degree, although no longer to the same extent as it used to do. The United States has $2\frac{2}{3}$ miles of railway per inhabitant against the United Kingdom's five-sixths of a mile, and although especially in this instance, and because of the peculiar conditions prevailing in the States, "comparisons are odious," one may from these figures deduce that competition is keener across the Atlantic than here. How keen it is may be seen from a few facts like these: Chicago and St. Louis have an aggregate population of 1,750,000, and the State of Illinois, through which nearly all connecting lines pass, boasts of 4,000,000 inhabitants, those of Chicago of course included; seven railway companies cater for traffic between the two points. There are four great direct lines between such points as Omaha and Denver, five between Chicago and Cincinnati, six between St. Paul and Kansas City, and seven between Chicago and Des Moines (Iowa), while as many railways connect the Western metropolis with St. Paul and with Peoria. These numbers only relate to direct routes, and if we counted all intermediate lines, most of them would be considerably increased. For instance there are twenty-two different "lines" between Chicago and New York, and between the latter city and New Orleans there are not less than a hundred and six different routes along which freight can be and is shipped almost daily, although the distances vary from the minimum of 1,180 miles, to the maximum of 2,053. The rates are in all cases virtually the same for through shipments; otherwise these "lines" would not be competitive.

As to the causes of this competition, although no very great practical importance attaches to them, it may be useful to state that the considerable profits to be made out of construction, in addition to that sanguine disposition of Americans which induces them to "overdo" so many things, were the most potent. And as to consequences, they all led to that "downward tendency of rates" so well known to the disappointed investor.

Let us briefly state what remedies were applied to cure this malady, and with what results. We have already said that the disease originated in excessive construction, and its symptoms in all cases consisted of financial debility, sometimes seen in its chronic and sometimes in its acute form. The first medicine administered was decidedly homœopathic; apparently it was thought that the fever would kill itself—that the weak would succumb and the strong survive. But expectations in that direction led to disappointment, because the weak possessed a considerably greater amount of vitality than they were credited with, and a general demoralisation of rates followed. In many instances this led to the destruction of the weak lines, and their subsequent absorption by the strong; rate wars destroyed the feeblest first and strengthened the hands of the others; but this victory was bought very dearly, and it soon became evident that a cure could be expected sooner from harmony than from war—and hence harmony was aimed at. It is not necessary to trace its establishment through all stages; suffice it to say that the endeavours led to the more or less universal introduction of "pooling," which was generally regarded and defined as a means of allaying competition.

In theory, "pooling" is no bad thing; it is applied successfully and without serious injury to the public in many parts of Europe, but in America it has been a failure, and it was this because of the lack of good faith among the railways. "Pools" were simply made to be broken. Every railway violated them, secretly and dishonestly to steal a march upon its rivals, and openly for speculative purposes. The Achilles' heel of "pools" is that adhesion of all their members is necessary to give them strength, and that dissent of one is sufficient to impart weakness; and this proved fatal to all. In the days of pools, honesty among American railroad managers was still at a discount, and most of them patronised Wall Street; and if a "bear" had the power to break up a pool, the temptation usually proved too much for him. Rate wars never were as frequent as in the days of pools, and this is why pools never were successful. A week of cutting would off-set a year of harmony, and since pools were detrimental to trade on account of the uncertainty of transportation charges, they were injurious both to the corporations and the community at large.* Hence their prohibition by the Inter-State Commerce Act, all that has been said to the contrary notwithstanding, was a distinct advantage to all concerned. It is equally certain that pools in no way stopped the decline in rates, which fell to such a phenomenal extent. From about five cents. in the early days of railroading, they dropped to 0.93 cents. in 1891.

Since rate wards and pools failed to stay this decline, we must see whether other remedies could be devised, and if so, in how far they were efficient. It is gratifying to find that there was not only a remedy, but also that it consisted, not of artificial ingredients, but of natural components. To employ as few words as possible, the low rates were offset by consolidation and by technical improvements. Rate wars, in other words low rates, exhausted many a system and caused it to be absorbed by stronger ones; in addition, the constant struggles tended to promote a combination of forces, and hence originated that consolidation which has been the most conspicuous feature of American railways. To describe its extent would be superfluous here, but it may be useful to mention that within the past twenty years more than 2,500 companies have been merged into others, and that to-day twenty-five corporations directly or indirectly control over 100,000 miles of railway.

---

* We return to the subject of pooling in Chap. IV.

The results of consolidation are obvious. There is, in the first place, that immense saving of energy always attending the amalgamation of sundry small concerns, and the economy resulting from it; in the second place, the combatants become more powerful, and war more destructive, and therefore more dreaded. Just as the growth of empires diminished the number of wars, growth of railway systems promoted peace; and consolidation advanced, not only because war is more destructive, but also because peace is now recognised to be the only remedy for low rates, and the natural antidote against destructive competition. The acquisition of control through ownership, leases, and agreements, has furthered an *entente cordiale*; and, apart from these, traffic associations, born of a desire for harmony, have promoted that amicable understanding which is essential to the profitable working of railways in a country where they compete vigorously.

But in spite of all its salutary consequences, consolidation alone would not have sufficed to offset excessive competition; rates fell too much for that. The American railway charges less to-day for carrying freight than it costs railways in other countries to move it; on English railways, for example, the cost price of moving freight is, as a rule, at least twenty per cent. higher than the selling price of the same amount of transportation on the average American railway. Yet, in spite of the fabulously low level of rates, it was absolutely necessary to reduce the cost of transportation to such extent that business would become profitable: without such reduction the entire interest would have been doomed to ruin.

This problem has been the most difficult American railroads were ever confronted with, and it must be admitted that it has been splendidly solved. American railways contrive to make a fair profit with rates that would be ruinous to the railway companies of any other country. By gradually introducing an astounding technical perfection they have saved themselves from financial ruin. During twenty years which saw relatively little progress in England, American railways advanced with rapid strides, and although our engineers for the greater part are still loth to confess it, the standard American railway of to-day is ahead of ours, and even the worst line can, in some respects, teach us many a valuable lesson. Our English railways are guilty of preventable and wanton waste, and not least of waste of motive power. In this country, the cost of moving freight is much higher than it ought to be. There is not a single valid reason why, with our superior engineers, our mineral wealth, and our cheaper labour, it should be higher than in America, where it has been reduced to a minimum by the force of necessity.

It is not necessary to say much of passenger traffic, since to most American lines this is of secondary importance, although it may be observed that, in proportion to its cost, passenger accommodation in America is far better than here; while even in the matter of speed, the best trans-Atlantic railways beat the cis-Atlantic.* Passenger business in America is much of an advertisement, and rarely very profitable: the movement of freight is the pivot around which the business of all but a few companies turns. And in this branch of business small wonders were worked in a very few years. To begin with, the roadbeds of most lines have been wonderfully improved; the majority are now well graded and ballasted, and, except on sidings and unimportant feeders, steel rails are universally met with. The rapid substitution of steel for iron tracks alone is a marvel. In 1880 there were 33,680 miles of steel rails against 81,967 of iron; in 1890, 167,606 miles were of the better metal, and only 40,697 remained of the baser; and everybody who knows how much the capacity of a steel track exceeds that of an iron, will understand what economies this change alone rendered possible. Still, the increase in the size of cars was much more important than the rapid introduction of steel rails for iron. Large wagons, of 60,000 lbs. capacity, are in universal use; and the amount of dead-weight has been reduced to a minimum. In 1870, a Pennsylvania standard car weighed 20,500 lbs., and its carrying capacity was 20,000 lbs., so that to move a ton of paying freight it

---

* The Empire State express, on the New York Central, daily runs from New York to Buffalo, 436½ miles, at a speed of 52¼ miles an hour, inclusive of stops. The "World's Fair Flyer," running along the same road and the Lake Shore, covers the distance between New York and Chicago in 20 hours, and attains an average net speed of 51 miles an hour. It easily holds the world's record for fast long distance runs.

was necessary to haul rather more than a ton of dead weight. To-day the American car of the standard type weighs twelve tons and carries thirty.

In England, and in Europe generally, a freight car of the most approved type weighs five tons, and carries eight; which means that the proportion of dead weight to paying freight in America is $1:2\frac{1}{2}$, in England only $1:1\frac{3}{5}$. In addition to this an American train usually carries as much as the condition of the way will permit, while in England ridiculously short trains can be seen. This fact, like the greater speed of freight trains in England, is no doubt connected with the more exacting demands of the public here, but there can be little doubt that expenses could be reduced very considerably if only serious attempts were made. The greater capacity of cars, the heavier train-loads, and the slower rate of speed, account for the lower cost of transportation, and, in addition, the locomotives are stronger and use a trifle less fuel than ours, while, above all, much more service is got out of rolling stock, cars being made to to perform twice as much work as in this country. All this shows American railways to be in many ways superior to ours, and as improvements continue at a swift pace, whilst we are almost at a standstill, the likelihood is that we shall be left behind more and more as time goes on.

In connection with these technical features it is natural that we should confront ourselves with the question, "Will proportionate profits increase or not?"

To this I believe an affirmative reply must be given, though it will be necessary to qualify the answer. The fall in rates seems to have reached its limit; charges no longer tend continuously downwards, but oscillate around a point approaching the average for the past six years; moreover, competition is gradually becoming less destructive, owing to the wholesome influence exercised by the introduction of more sensible methods—methods which find expression in the establishment of Associations and also in consolidation; and although rates in some sections will no doubt require further adjustment, they will presumably rise in those regions where local traffic is on the increase, and there will in all probability be no further serious fall, and no renewal of rate wars. To this rate question I propose, however, to return at some length later on.

Whilst rates are fairly approaching "bed rock," the expenditure involved in operating the various properties becomes cheaper as time goes on. Perfection of roadbeds and rolling stock, as I have said, continues; and the higher the level it reaches, the cheaper freight can be moved. What economies arise from this cause alone may be inferred from the fact that the Lake Shore Railway, one of the best lines in the country, does a more remunerative business than the Wabash (which as far as geographical situation is concerned is its equal), although its rates are lower, and this solely because it can move a ton one mile at a cost of 0.458c., whereas its rival, just as favourably situated, but possessing an imperfect roadbed, cannot do it for less than 0.563c. This simply means that the Wabash, were its road as perfect as that of the Lake Shore, would earn over one and a quarter million dollars per annum more than it does now, and from this the reader may deduce what importance attaches to technical perfections; at the same time he can see how foolish it is to object to constant betterments; no matter whether paid for out of earnings or charged to capital, all judicious expenditure upon them must prove true economy in the end.

Improvements being effected everywhere, it follows that the proportion of operating expenses to earnings must grow smaller, and as rates will not be subject to a further appreciable fall, it seems certain that the movement of freight will gradually become more profitable. The one event which could prevent this would be a resumption of reckless construction; but for various reasons this is not likely. The mere construction of a railway is no longer a profitable business, because the immense systems of to-day are not coyed or cowed into "buying out" as easily as were the small lines of the past, and in addition illicit gains have become well nigh impossible because of the higher standard of morality now prevailing throughout the country, and of the wholesome influence of inevasible publicity. To this last fact I point because it is well known that the profits which could be made by the aid of "watering," construction companies, etc., were one of the principal causes of that injurious disproportion between the supply of transportation and the demand for it which was the principal, if not the only cause, of the fall in rates.

The improvements in the relations between the railways and the Republic, and in the prospects connected with rates, do not constitute the only amelioration that has to be recorded; those with the investor have also undergone a salutary change. But this matter may not be classified with those dealt with in these introductory and more or less prefunctory pages. It requires chapters of its own.

CHAPTER II.

AMERICAN RAILROAD FINANCE.

We have just dealt with railroad matters in general, and outlined the relations existing between the companies and the public, and between the railroads amongst themselves. We now have to consider the *entente* between the corporations and the investor.

Before we enter upon this vast theme it is, however, desirable to clearly define the fundamental principles guiding these relations. These principles differ radically from those which underlie railroad finance in other countries, notably in England; and it is essential to a thorough understanding of the whole subject that the difference should be most clearly realised. To impress its existence upon the reader's mind is, indeed, the principal purpose of this book; and for that reason I shall devote a separate chapter to an exposition of the principal distinctive features of American railroad finance.

The initial chapter must have convinced the reader of the existence of marked differences between the American railroad and its English sister; and indeed the two have but little in common. Even cars and couplers, rails and road-beds, brakes and bridges are entirely different in the two countries; the circumstances amongst which the railways came into existence and their present surroundings present but few analogies; their relations with the people and with their rivals are not in America as they are "over here." But nowhere are differences so striking as in matters pertaining to finance; indeed, in this respect one can hardly speak of differences; there are contrasts, and contrasts of the most striking kind.

A little reflection will show that the existence of dissimilarities between American and English railways is quite natural. The two countries and their inhabitants differ widely from each other; and social institutions being always indelibly impressed with the stamp of the surroundings and social tendencies of which they are the creatures, it follows that the essential characteristics of railways must differ in various countries, and harmonise with the most prominent features of the nations amongst whom they exist. Thus we find that the English railways are entirely in concord with the thoroughness and the conservatism of the English people; in America on the other hand the railroads remind us of the energy and enterprise of the Yankee, of his occasional lack of thoroughness, of the mushroom growth of his country, and so forth.

In but few respects is the contrast so striking as with regard to the relations between the railroads and the law. In England the collective wisdom of Parliament has to decide whether the needs of the public call for a new line or not, and has to approve not only of the project but also of the cost of a new railway. The smallest issue of capital must be sanctioned by Act of Parliament, and the authorities look after many details of service, prescribe the introduction of safety appliances, and regulate the working hours of railway employées, whilst maximum tariffs are laid down within the bounds of which the companies are compelled to keep. In the United States on the other hand legislative supervision is for the most part absent. There are a few feeble enactments, passed after decades of abuses, which serve to protect the people against downright tyranny or extortion;* but for the rest every one, so to speak, can construct a railway wherever he likes, and can build and work it in whatever manner he pleases. He can issue as much stock as he sees fit, introduce as few or as many safety appliances as he thinks desirable,† charge whatever rates he deems best, and regulate the working hours of his employées at his own pleasure. Thus we find that the American railroad enjoys a much higher degree of freedom than the English; but in no respect, perhaps, is it so absolutely free from any kind of legislative interference as in matters pertaining to its finances.

This lack of Government control exercised a most direct and decisive influence upon American rail-

---

* Amongst these I include the more or less transitory enactments of the State Legislatures, notably those attempting to fix tariffs. Reference is made to them in Chap. IV.

† Except in some of the New England States; but the railways in that section have no direct interest for the British investor. In New York State there is also some slight supervision over their finances.

road finance; in fact, it determined its fundamental principles. It also affected railroad development in general to a much greater extent than is usually supposed; for example, the destructive competition still prevailing in many sections of the country would never have come into existence if the American Government had interfered as much with railway enterprise as the British. But it is not necessary to refer again to the influence which the passive attitude of the Government exercised upon railway development in general; our present purpose is to trace and illustrate its immense bearing upon railroad finance, and to give an outline of those far-reaching consequences to which we shall frequently have to revert in the course of subsequent chapters.

It is obvious that the absence of Government supervision over railway finance left railroad directors in possession of an almost unlimited amount of discretion and power, and so to speak rendered them responsible to no one, and that it left the investor without any protection whatsoever, save that which he gradually succeeded in craeting himself. Hence, during the early days of railways, it was within the power of the managers to commit the grossest abuses, and this implies of course that the investor was, in a very large measure, at the mercy of those who managed his property.

This state of affairs could not fail to permanently affect both principles and methods of American railroad finance. It rendered the investor helpless,

and therefore cautious and exacting; it made the
manager omnipotent, and consequently independent,
irresponsible, and often dishonest; and by doing this
it determined the attitude of both, and decided their
mutual relations. More than that; every one of the
peculiar characteristics of American railroad finance,
all its evils and peculiarities, spring from this absence
of efficient supervision over capitalisation.

As far as the investor was concerned, this fundamental defect, it is almost superfluous to say, necessarily rendered him cautious and circumspect. There
being nobody who undertook to protect him, the
capitalist was bound to protect himself; there being
no enactments which looked after the custodians and
managers of his property, it was incumbent upon
him to give those managers as little power as he
possibly could. And, therefore, instead of becoming a shareholder of the new railways, instead
of figuring as a proprietor who was to share all
profits, but who had also to run the risks and to bear
the losses, the investor contented himself with being a
creditor who renounced his share of possible profits
in order to avoid possible risks, and who preferred
advancing money secured by a lien upon the property
to investing in the property itself.

From this attitude there sprang one of the most
striking characteristics of American railroad finance.
The railways of nearly every other country have
been built exclusively by shareholders, those of the
United States were constructed mainly with the pro-

ceeds from the sales of mortgage bonds. *Funded debt became the basis of American railroad capitalisation.*

Whilst it left the investor cautious and helpless, the absence of efficient supervision over capitalisation rendered the railroad "boss" audacious and powerful; and this irresponsibility and independence was fraught with danger in a community which twenty or thirty years ago was absorbed in a feverish scramble for wealth, and at a time when business morality was at a discount. And if we recall to our minds the condition of business life during the era of railroad construction, we can only find it natural that those who constructed and conducted railways went as far as their unrestricted powers permitted them to go, and furthered their personal aims and ends as much as they could without the slightest regard for the interest of the property, its owners, or its creditors. Thus the passive attitude which the Government saw fit to adopt with regard to railroad capitalisation may be held directly responsible for those malpractices to which we shall refer at some length in subsequent chapters.

But time, which has wrought so many quick changes in the affairs of American railroads, has also reformed their financial methods. True, in matters of railroad finance, the Government adheres to the policy of *laissez faire* it has always pursued; but other circumstances have fortunately tended to mitigate the results of this neglect to a very considerable extent.

It is quite evident that the old conditions were little calculated to promote confidential relations between the managers and the capitalist. The former were endowed with vast powers which they abused yet endeavoured to keep intact; the latter suffered from these powers and therefore tried to check and curtail them. Hence the aims and wishes of the "railroad boss" and of the investor were conflicting, their attitude was more or less antagonistic, and they had little faith in each other. Both were engaged in a constant struggle for supremacy; but although this contest has continued until the present day, the capitalist has gained much ground, whilst there can be no doubt whatever that he will ultimately take possession of the field.

This quiet contest reminds us very much of the attitude which the English Parliament assumed towards the Sovereigns of old. When the Crown required money, the people's representatives often refused to give it unless it yielded some of its prerogatives or privileges; when an American railroad company appealed to the capitalist for funds, the latter frequently exacted stipulations aiming at a restriction of the powers of its managers. Every issue of bonds offered an opportunity, of which the investor could avail himself, to increase the restraint he placed upon the discretionary powers of the managers; these opportunities were usually taken advantage of, most appeals for new funds being granted only against the surrender of some part of

the manager's power. In this manner the investor
gradually succeeded in limiting the undesirable discretion
of managers. For example, he frequently curtailed
the power to issue new capital,* and often succeeded
in prescribing the application of revenue by
means of stipulations in mortgage deeds.

But the rude force of the investor's millions was
not the sole agency which furnished the check the
Government failed to provide. That the discretion
of railway directors is abused far less frequently now
than it used to be must also be attributed to the
improvement in business morality which has taken
place in the United States. The present generation of
railroad managers has a loftier perception of its duties
towards shareholders than its predecessor; in addition
the railroad manager of to-day has been taught
that his own interests demand that he should establish
the credit of his company on the soundest
possible basis, and this, as well as the amelioration
of business morality in general, were additional
remedies against the evils that sprang from the laxity
of the Government, and have assisted in removing its
most baneful results and in lessening the friction
between the manager and the capitalist. Of course we
are here speaking generally; there still remain some
railway managers who either care little about honesty,
or else fail to see that their great discretionary

---

* In many cases even the issue of common stock is limited now-a-days,
whereas formerly it was hardly restricted in any case. But
there is still room for considerable progress in this direction. An
issue of bonds is, of course, limited in all instances; at any rate, I
know of no case where it is not.

powers are antagonistic to the best interests of their employers. But speaking broadly it may be said that the managements of the vast majority of companies have ceased to strive after preservation or increase of undue discretionary powers, and that they recognise the great interest which their properties have in a restriction and clear definition of the discretion which must necessarily be vested in those who are intrusted with the management of great railway corporations.

With these general remarks we will for the present take leave of this subject; to its details we shall have to revert on several future occasions. Before we conclude this chapter it is, however, pertinent to call attention to two facts. Firstly, it is evident that the investor who appreciates prudence and safety should as a rule never invest in any funds the yield of which depends in any direct way upon the discretion of managers; even where there is no reason for distrust, it is best to remain on the safe side. The investor should confine himself exclusively to bonds, and at its best to preferred shares, the precedence of which is assured by stipulations preventing any further issue from taking place without his consent, and the returns upon which are in no way dependent upon the decisions of managers.* Secondly, the foregoing implies that the discretion of managers extends over two distinct matters: capitalisation and the applica-

---

* The returns on most preferred shares *must* be paid if earned, although it is possible to make this "if earned" a very elastic term, by spending earnings on betterments. [See Chap. V.]

tion of revenue. In how far it does this will be fully elucidated by the two following chapters, which deal with these intricate subjects. But when perusing these chapters the reader must bear in mind that the principles of American railroad finance are entirely different from the principles adopted in this country. It is essential that he should not lose sight of this cardinal point, that he should know why the difference exists, and that he should remember that different principles must needs lead to different methods.

CHAPTER III.

CAPITALISATION.

The peculiar relations between the investor and the corporations, to which we have just called attention, have had a host of consequences; but their influence upon capitalisation has been greater than their bearing upon any of the other affairs connected with railroads. Railway capital, as we have already had occasion to state, contains a fictitious element of considerable magnitude, commonly called water; and the origin of this "water" is intimately connected with the unrestricted powers which railroad managers had in the past.

There prevail a good many erroneous ideas concerning water. Neither its origin and nature nor its effects are as fully understood as they ought to be, and the result is that most people have no very clear ideas upon the subject. The common version is that the causes, characteristics, and consequences of water are equally baneful, and that it is an indelible taint left upon American railroads by an era full of gross abuses. But this view is by no means absolutely correct. The origin of a vast amount of water is no doubt as disreputable as it possibly could be, and its effects are in many cases highly injurious; but a considerable part of it is the result of necessity or expediency, and the effects are not nearly so bad as is generally supposed; in some respects they are even beneficial to the investor. There is, in fact, water and water; and to demonstrate the truth of this proposition we shall discuss the subject as fully as the scope of this little volume permits, and show its origin and consequences. This will be more easily effected by enumerating the various forms it took under the influence of the circumstances which called it into existence.*

There were in the main six different ways of inflating the capital of American railways, namely :—

1. By fraudulent issues of shares and bonds.
2. By paying too much for construction.
3. By purchasing properties at excessive prices.
4. By buying superfluous competing lines.
5. By selling bonds and shares at a discount.
6. By declaring stock dividends.

Fraudulent issues, of course, were the worst form of "watering." They were effected by the managers solely for the purpose of self-enrichment, either directly or else indirectly in connection with market manipulations. The Pacific and Reading railroads have in especial suffered very severely from this malpractice, but the Erie has probably been cursed with it more than any other company. Between 1868 and 1872 "Jim" Fisk and Jay Gould increased the share capital of that corporation from $17,000,000

* This chapter contains several pages transcribed, with but immaterial alterations and some additions, from *American Railroads as Investments*.

to $78,000,000, mainly to manipulate Wall Street, and President Watson doubled the funded debt a few years later, it is said also chiefly for his own benefit. At that time Wall Street was not exactly an abode of saints; but even for Wall Street these proceedings were too bad, and in 1869 the Board of the New York Stock Exchange refused to quote these shares any longer. But there are dozens of other companies which suffered to almost an equal extent, or at any rate, in the same proportion.

Hardly less repulsive than this practice was the second, consisting of paying excessive prices for construction. In the early days of railroading it was not unusual for railway companies to enter into contracts with construction companies. These noxious concerns were usually composed of members of the board and their friends, and of course served no honourable purpose, their object being to enable directors to make money at the cost of the investor. To show how the capital of companies was inflated by this means, I will quote the case of the now defunct South Pennsylvania Railroad Company, started by Vanderbilt to compete with Pennsylvania Railroad. This road has been proven to have cost not more than $6,500,000, and a responsible contractor had offered to build it for that sum. Yet a construction company, composed of Vanderbilt's clerks, received $15,000,000 to complete it; and the syndicate of capitalists which supplied this money got $40,000,000 in bonds and shares, so that for every dollar of actual cost over six dollars of bonds and shares were issued. In the same manner, though not in the same proportion, the thing was worked all over the Union, especially in connection with the Pacific roads, a group of railways which has seen more frauds than any other. The builders of the Central Pacific, for instance, commenced with the modest sum of $159,000, and taking this as a nucleus they completed the road, gathering a total capitalisation of $139,000,000, and acquiring large fortunes over the transaction; the Government Commission on Pacific Railroads in its Report to Congress says that $58,000,000 would have been a very good price for the railway. Car Trusts were—and in some cases still are—another means of robbing companies for the benefit of their managers, the shareholder being compelled to pay an excessive price for rolling stock supplied by the Car Trust on the instalment plan.*

On a par with construction companies was the purchase of other properties at excessive prices. The Coal and Iron Company of the Reading, for many years a great burden to that line, has been quoted as one of the most notorious cases, this concern being paid for, it is said, at the rate of at least twice its intrinsic value. Years ago it was a common thing for railroad directors to buy a property in their private capacity, in order to sell or lease it to their company

---

* One of these Car Trusts is connected with the Chicago, St. Paul and Kansas City, now the Chicago Great Western Railroad. I have tried hard to find out what it is like, but failed to get any other information beyond that contained in a letter from a North Western railroad resident, who tells me he is under the impression that the Car Trust is a parasite of the road.

at an immense profit; and until twelve or fifteen years ago the majority of purchases of auxiliary concerns were permeated with fraud. Many leases, too, had little in common with Cæsar's wife; the history of the Wabash bears out this statement. But I believe this kind of fraud is very rare now.

The purchase of "parallel" lines, although in numerous instances taken advantage of to carry out some "deal," was in a measure unavoidable, the independent existence of such lines being a constant danger to others, and destructive to all concerned. The Vanderbilt interest was practically forced to lease the West Shore and to buy the "Nickel Plate" unless they wished to see their other properties turned into financial wrecks; and the Pennsylvania had to arrive at a compromise with Vanderbilt, resulting in the abandonment of the construction of the South Pennsylvania Railroad, if it desired to maintain its position in Pennsylvania. On the whole it may be said that this form of watering, under the peculiar circumstances that prevailed, could not have been prevented. Yet it caused a considerable amount of superfluous capital to be employed in transportation, and in some degree the methods of managers were to blame for this, inasmuch as the existence of a line overburdened with water tempted to the construction of another which was free from this great impediment, and could therefore work at rates which would have been ruinous to the other company.

These four were the noxious forms of water, and

we have no desire to say a single word in defence of them. But the amount of water resulting from these malpractices is only a part of the whole; and the remainder springs, not from dishonesty, but from necessity. Take, for example, the fifth of the forms enumerated above—selling stock and bonds at a discount. This was the most widespread cause of water; yet but few objections can be raised against it, and it was excusable in almost every case. In the early days of railroads especially there was little inducement to invest money in these enterprises, and to create such inducement bonds had to be sold at a discount, and shares were frequently given into the bargain. But even Governments sell their securities at a discount if they cannot get rid of them at par or at a premium, and the railways, by their prospects, were justified in borrowing money at excessive rates if they could not get it otherwise. Like the cotton planters of the South, who after the war often paid fifty per cent. interest on the capital required for bringing their cotton land under cultivation, the railways would have outgrown the payment of excessive rates for money if their affairs had otherwise been conducted with honesty and integrity. Unfortunately, at a certain period these two qualities, instead of being the rule, were the exception; and as we have seen above, those who did not hesitate to pay dearly for the money they required, instead of endeavouring to effect economy in other directions, evidently thought that, a heavy handicap having been placed upon the railways in the shape of dear capital, it could not matter much whether the burden was increased a little or not. But this fact furnishes no argument against the sales of securities at a discount as long as these sales were effected on straightforward lines.*

But the abuses of the early days of railroads, above referred to, and the sale of securities at a discount, were not the only causes of "water"; the fictitious capital resulting from the payment of stock dividends also reaches formidable dimensions. The most famous stock dividend ever distributed was one of 80 per cent., paid in December, 1868, on the shares of the New York Central Railroad Company, and eleven months later when the consolidation with the Hudson River Railroad followed, a further stock dividend of 27 per cent. was declared, while the Hudson River Railroad shareholders received one of 85. The Reading paid a scrip dividend of 10 per

---

* The practice of selling merely bonds, and giving shares with them, certainly looks peculiar. But it should be remembered that nobody felt great inclination to advance money unless it was upon security, and that it was usually impossible to find a sufficient number of *bonâ fide* shareholders. Many people deemed even bonds no sufficient equivalent for their investment, and besides prior rights upon the property they also wished to possess control of the company, and a chance of partaking in the future profits; and shares being not much sought after anyhow it mattered little to the promoter whether he gave them into the bargain with bonds or not. The majority of companies realised nothing for the shares they issued in their early days. The Missouri, Kansas and Texas Railway Company, for instance, gave $21,400,000 in shares to a construction company, in addition to the payment made in bonds. The New York Central, Erie, Reading, St. Paul, Chicago and North Western, in short, almost every railway company, as a rule parted with the earlier issues of its common shares without getting any cash for them, although, as is well known, shares are no longer given away now-a-days, and frequently sell at good prices.

cent. in 1846, and one of 12 in 1847, while between 1871 and 1876 upon a capital of $34,200,000, already fictitious to the extent of probably more than one-half, $15,700,000 was paid in dividends, mostly in scrip. The Erie Railroad made still larger payments of stock dividends; the Chicago, Burlington and Quincy Railroad paid 20 per cent. in 1880, the Atchison, Topeka and Santa Fé 50 per cent. in 1881, and the practice may be said to have been general, and is still resorted to in numerous cases. Only recently the Pennsylvania distributed a small scrip dividend, which seems to have been declared chiefly for the purpose of sustaining the quotation of its shares.

None of the six methods of watering enumerated and described above call for special comment, except the last. A stock dividend is somewhat of a farce, and of a deceptive nature. Several objections have been offered against it; and on the other hand it has been defended with forcible arguments.

Scrip dividends are frequently paid to shareholders as a substitute for cash dividends which have been passed. A board feels that shareholders are entitled to some return upon their capital; earnings have been employed to pay for betterments, and having no cash they pay scrip or shares. A moment's reflection will show that such a dividend is really no dividend at all. It creates more shares, and hence future distributions must be so much smaller. If a company pays a stock dividend of fifty per cent. its future

dividends must be 33⅓ per cent. less than otherwise,
and hence a scrip dividend, looked upon as a return
upon capital, is delusive. Whosoever receives it does
not actually get a farthing more upon his investment
when cash dividends are resumed.* Their deceptive
nature therefore leads to the conclusion that other
reasons exist for their payment, unless it is a mere
sham, and, indeed, their distribution is based upon
three circumstances—the rise in the value of railway
property, the payments made for betterments out
of earnings, and most of all, the tendency of Legislatures to cut down rates whenever they are so high
as to make the possession of railway stock remunerative. The American people, although on the whole,
I think, by no means unreasonable, are opposed to
large corporations, a fact so abundantly demonstrated
by the attitude of their Legislatures and Courts of
Justice that it is not necessary to supply proof of the
assertion. They do not stop to think that action

---

* Yet scrip dividends do not altogether fail to benefit shareholders as a mass. Experience has shown that a stock dividend usually enhances the prices of shares; for instance, if a dividend of, say, 50 per cent. is declared upon a stock quoting at 60, the price of the latter will rarely fall to 40 as it ought to do, but as a rule remains above that figure, so that usually there is a gain for its owner, which, it is asserted, offsets the non-payment of dividends. But with regard to this matter the fact should not be lost sight of that these profits rarely reach the persons for whom they are intended, namely the investors who perhaps for years held an un-remunerative security. Shares change hands more frequently than bonds, and hence the benefit of stock dividends usually falls upon people who have not deserved it by former sacrifices. But the fact that stock dividends proportionately raise the prices of shares renders them especially objectionable, inasmuch as it constitutes a temptation for directors to declare them and to speculate. In-stances are on record in which such dividends were declared solely for speculative purposes.

detrimental to a vast interest like railways must fall back upon the people; they consider anything done to its disadvantage to be a public gain. Hence the people are "hard on railways," and as the railways by their conduct hardly promoted an amicable *entente* with the people, (see p. 9), the latter had a good deal of provocation, and often became unreasonable. For instance, among the people of the North-West it is almost a maxim that no railway should be allowed to pay more than four per cent. on its common stock. " The stock being water to the extent of at least one-half," the Granger reasons, "four per cent. means eight, and eight will do." But those who talk in this strain overlook two important matters. In the first place they forget that the shareholder, no matter whether foreign or negative, has for many years held unremunerative stock, that in bygone days he has incurred risks and suffered losses, and that hence he is entitled now to returns which offset these. In the second place they do not recognise the principle that a railway company, as well as a private individual, is entitled to reap the benefits of an increase in the value of its property, and may lay claim to compensation for the risk run by early investments to which, more than to anything else, the wealth and the welfare of the Great Republic are due.

This very unreasonable attitude compelled the railways to ward off its effects; and for that reason they frequently resorted to stock dividends as a dodge. By inflating their share capital, the companies endeavoured to conceal profits, in order to thwart the designs of rate-reducing Railroad Commissioners; and this was necessary because the shareholder was entitled to compensation for the risks incurred and the losses suffered in the past.

We have just stated the causes of water; it now remains for us to deal with its extent and consequences.

The foregoing remarks must have given the reader the impression that a formidable proportion of the capital now nominally represented by American railways is fictitious. Such is indeed the case. It is, for example, generally conceded that the New York Central, Erie, and Reading companies between them have water to the amount of at least $200,000,000, while some, Mr. J. F. Hudson for instance, estimate it at $300,000,000 out of an aggregate capitalisation of $500,000,000; and from this we can infer how much water the entire system contains. Both in respect of the extent to which watering has been carried on, and the degree in which the process has been applied, we find some interesting details in "Poor's Manual" for 1884. In that year Mr. H. V. Poor wrote:—" The increase of share capital and indebtedness of all companies for the three years ending December 31, 1883, was $2,093,433,054, the cost of the new mileage as represented by share capital and debt being about $70,000 per mile. The increase in the three years of the funded debts of all companies was $924,165,440;

of their floating debts, $169,880,406; of the two $1,094,045,846. It is not probable that the cost of the mileage constructed in the three years equalled the increase of funded and floating debts by at least the sum of $200,000,000. The cost of the mileage constructed certainly did not exceed $30,000 to the mile. The whole increase of the share capital, $999,387,208, and a portion of the funded debt, was in excess of cost of construction. It will be seen by a statement hereto annexed that stocks and bonds to the amount of $530,132,000 were listed at the New York Stock Exchange in 1883. The amount of stocks and bonds listed was equivalent to about $80,000 per mile of new road built during the year."

From this it seems we may infer that the average cost of American railways in 1883 was about $30,000 per mile; and if, on account of subsequent betterments we make a further allowance of $10,000, we find that to-day the total capitalisation ought to be $6,840,000,000 for 171,000 miles, whereas it actually reaches $10,122,500,000. This estimate, however, is in favour of the railways, for Mr. Poor, in his Manual for 1884, further says that the *bonâ fide* investment in railways probably does not exceed the aggregate of their floating and funded debts, and if this holds good to-day there would be, not as I presumed above, $3,282,000,000 of water, but $4,640,000,000. Some writers even go so far as to allege that the estimate of Mr. Poor, who is deemed a spokesman of the railways, is moderate and conservative, and the fictitious capital is said by some to amount perhaps to fully two-thirds of the total.

The consequences of "water" are as varied as the means by which it was effected. Like most railway questions they admit of divergent views, varying according to the position one takes; it is, for example, evident that the public sees the matter in a different light than the companies or the investor. The principal argument on the side of the public is that "water" engenders a desire to charge such rates as will result in good returns upon capital; this is an undisputed and an indisputable fact, although the railways seem to consider it bad policy to avow it frankly. Water led to pools; it resulted in discriminations which were applied because it was expected that they would enhance earnings; and it induced the railways to fix such rates as would enable them to pay good returns upon an inflated capitalisation. In view of facts like these it is natural that the public, anyhow clamouring for low rates, and unwilling to enable railroads to pay heavy returns upon their capital, strongly denounced watering.

As to the investor, until a few years ago he had as little reason to approve of watering as the public. In the first place, water was largely the result of countless frauds, and the possibility of creating it was responsible for numerous malpractices. Secondly, it had been the cause of competition; wherever a line burdened with "capital" could be found there existed the temptation to build another with less stock, which, since it had to earn but one-half or one-third of the amount required by the existing company to pay the same return per cent. upon its small capital, could enter into a very successful competition. In the third place, there would have been fewer rate wars, not only because, as we have just shown, there would have been less competition, but also because all companies would have been placed on an equal basis —that of actual cost—in their endeavours to earn fair returns, with the result that "cuts" would have been put a limit to by the requirements of solvent systems. In the fourth place, if there had been less water there would have been better returns upon the capital, bonds could presumably have been placed at four or five per cent. instead of seven or eight, and the returns upon shares would have been higher; but as a low rate of interest is scarcely desirable from the investor's point of view, this aspect may perhaps be left alone.

This being the true nature of water, the fact that it engendered a distrust which dealt many heavy blows to the credit of American railroad securities is by no means surprising. To quote once more from Mr. Poor's Handbook (for 1884):—"It is in this (immense " increase of fictitious capital) that is to be found the " cause of the general distrust which prevails, and the " enormous decline in the price of railroad securities. " From 1879 to near the close of 1883 a most singular " delusion rested upon the public as to their value, and " this delusion was taken advantage of on a vast scale

"by able and unscrupulous adventurers. Whatever was
"manufacturered and put afloat was seized with avidity
"by an eager and uninformed public. The delusion
"was increased and prolonged by payments on a very
"large scale of interest and dividends from capital. In
"this delusion the most loud-mouthed and unscrupu-
"lous promoters usually had the greatest success. The
"delusion culminated about the time of the opening of
"the Northern Pacific, in connection with which vision-
"ary schemes of immense magnitude had been put
"upon the market. Their worthlessness, and the rapid
"decline of their securities, exerted a powerful influence
"over the public mind which continues unchecked to
"the time of writing this. The distrust extends alike
"to good and bad, so that prices at the present time
"have as little reference to values as they had at the
"beginning of 1883. The distrust will probably con-
"tinue until time shall show what securities are really
"well based."

In England the distrust which sprang from the objectionable kinds of water lasted considerably longer, and in addition no allowance was made for those descriptions which were either unavoidable or else in the interest of the investor. Even to-day it has by no means disappeared, and with the old rascalities still lingering in the minds of our investors, and with an occasional fresh "scandal" to revive the distrust, this can cause little surprise. As recently as 1885 the public—it may be useful to say the American public as well—were imposed upon in a manner which

renders it difficult to say which of the two was
the more astonishing, the audacity of the impostor or
the credulity of those imposed upon. The exposure
of the swindles was followed by a distrust which so
entirely possessed itself of the public mind that it
lasted long after the cause had almost entirely disappeared, and in a measure it continues despite the
the great improvement in railroad matters and railroad morality which set in in 1885 and has continued
unabated ever since. To-day there may still be
several black sheep among the great flock of American
railways. But it cannot be questioned that both
the efficiency and the honesty of the management
have, on the whole, increased at a very rapid pace, and
that most of the companies have been weaned from
the bad practices which rightly brought all of them
into disrepute. The improvement has been altogether
marvellous, and in a measure the investor seems to
recognise it; at least this must be deduced from the
fact that issues of American railroad bonds are
eagerly purchased to-day at rates which ten years
ago would have attracted few investors. True,
this improved credit is not solely a consequence of
greater faith in these securities: the redemption of
United States Bonds, the conversion of the National
Debt, the increase in the number of Trusts, and other
developments have improved the demand for American
railroad securities, and probably contributed more to
their popularity than the consciousness that these
investments deserve greater confidence now than

they did ten years ago. But however much allowance must be made for these influences, it re-remains a hard fact that the investor's estimate of the merits of American securities has improved and keeps on improving.* Yet in England, and in Europe generally, there is not that faith in American railway stocks, and especially in American railroad bonds, which they deserve, and perhaps this lack of confidence is not remarkable. As recently as ten years ago methods of management gave abundant reason for distrust. We had been deceived and disappointed, and lost faith in consequence; and not being accustomed to observe rapid changes in our own surroundings, were slow to believe that in America a change from bad to good could be effected within a single decade. Yet this transition has undoubtedly taken place. A visit to the States, or a comparison of the conditions of to-day with those of the past, will readily convince even a sceptic, although it is far from us to assert that changes, however favourable, amount to a complete cure. Of so great a flock of scurvy sheep every single one cannot reasonably be expected to have become sound within so short a time. But although the order of things has not been absolutely and completely reversed, there are so many favourable changes that by this time the healthy element preponderates by far, and is daily growing in extent. In England most people do not appear to be aware of

---

* Of course we must look at this matter in a broad spirit, and disregard temporary depressions of the market, such as the one through which we are passing now.

this enormous alteration, and although there are abundant reasons for a revision of opinion, the same notions concerning "Yankee Rails" which were held ten years ago prevail to-day among the majority of European investors, who, unaware as they are of the great change for the better, still believe that American railway securities deserve little confidence and no reliance; while many continue to regard the companies chiefly as tools provided for the use of cunning men, and designed to fleece the foreign lamb. The reasons for this distrust have almost entirely disappeared, and the distrust itself is therefore no longer warranted. But to this matter we will return later on.

With regard to "water," people have modified their ideas as little as in respect of other matters pertaining to American railway finance. Ten or fifteen years ago it was said that nothing was more injurious to the investor than "water" and the frauds connected with it, and this opinion was not devoid of foundation then. To-day the same views prevail, although they are no longer supported by conditions. Some railway companies no doubt continue to suffer from the effects of water fraudulently poured out; but time has discounted its effects in numerous cases, whilst in others its influence has been to a great extent neutralised.

Water, regarded in its entirety, is not nearly as detrimental to the companies now as it used to be. Circumstances like the growth of the country, the expansion of systems and appreciation of properties have supplied a counterpoise for the fraudulent part of it, as they did many years ago in England ; and what we have termed the necessary portion, apart from being inevitable, is gradually becoming harmless instead of baneful. In a way it is even beneficial. All that has been said to the contrary notwithstanding, water emphatically begets a desire on the part of railways to charge such rates as will enable them to pay good returns upon a highly inflated capital. Whether they will in all cases be able to give effect to this desire is an open question; that they have succeeded in many is beyond dispute. Numerous companies pay good dividends upon largely inflated stock; others are trying to. Now let us suppose there were no "water." Then the companies, with the same earnings, would pay enormous dividends, probably an average of 10 per cent., and perhaps more. Such high returns of course could not fail to attract attention, and would undoubtedly invite new competition; but apart from this, would the nation with its never-ceasing complaints of rates—would the State Legislatures, which reduce transportation charges as soon as returns upon capital exceed a moderate figure—permit a return of 10 per cent. to be paid upon railway capital? Surely not. Hence it may be said that water is in a measure beneficial to the investor, because it is instrumental in giving him those high returns upon his *actual* investment, and those rewards for past losses and sacrifices, which would be impossible without it. And it is but fair that it should do so. Stock was given away and "water" created largely—though by no means exclusively—to give something which in the future would reward those who sacrificed in the past. Hence the investor of to-day receives great benefits from the wrongs inflicted upon his predecessor. He receives a good rate of interest on bonds which he bought at a discount: he gets fair returns upon shares which originally cost him nothing or next to nothing; the value of his investment in American railways has risen far above the price originally paid for it; and all this has been effected chiefly by means of the much-abused "water." Does that not imply that the prevailing ideas of water stand in urgent need of revision?

CHAPTER IV.

RATES AND REVENUE.

The initial chapter has already dealt with several aspects of the rate question, at some length with the causes, extent, and consequences of discrimination, and in a cursory manner with the constant decline in transportation charges. But this decline has not received that exhaustive treatment to which its importance entitles it. Rates, as everybody knows, constitute the only source of all railroad revenue; upon their adequacy depends the adequacy of the returns upon the capital invested in railways; and the subject, therefore, interests investors to a sufficient extent to render a full discussion desirable.

We shall not return to the complex question of discrimination. There cannot be any doubt that differential rates continue to exist in more than one guise, despite the enactments of the Legislatures; but the strenuous opposition offered to the pernicious practice has gradually suppressed its most vicious forms and reduced it to something like harmless dimensions. Railway economists of undoubted standing give us the assurance that railways cannot be efficiently worked without judicious application of differential rates, and as far as American railways are concerned we are not disposed to dispute this assertion, since the United States has little to complain of with regard to its rates which, taken all round, are far below those charged in any other country.* The question of discrimination does not, therefore, at this moment affect the position of American railways to an extent which invests the subject with supreme importance to investors. The railways themselves, and laws like the Inter-State Act, have reduced it to the smallest dimensions to which it could possibly be confined; and with competition rampant in all sections of the country, with the true principles of railroad management becoming more clearly recognised as time rolls on, and with the business morality of railroad officials subject to a rapid process of purification, discrimination no longer

---

* I published the subjoined table in the *Economist* of January 28th, 1893. The figures are all taken from official sources, whose accuracy cannot be doubted :—

		L. and N. W. Maximum rate per ton, 258 miles.	Dutch Tariff per ton, 395 kilom.= 258 miles.	Trunk Line rate per ton, about 950 miles.
1. Lowest class of freight	Class B in England ,, 6 ,, U.S. ,, C ,, Holland	s. d. 18 6	s. d. 8 4	s. d. 21 11
2. Middle class of freight	Class 3 in England ,, 3 ,, U.S. Piece goods, slow freight in Holland	53 1	15 4	43 9
3. Highest class of freight.	Class 5 in England ,, 1 ,, U.S. Piece goods, fast freight in Holland	76 5	30 8	65 7

D

exists in forms injurious to the interests of the commercial community.

Discrimination, therefore, has ceased to constitute the backbone of the rate question in the United States. The gross injustices of a dozen years ago are no longer committed. The undue preference shown to individuals, to localities, and to particular trades is a thing of the past, and speaking broadly all shippers now find themselves on a more or less equal basis. But instead of the hostile legislation engendered by abominable malpractices, the railways now have to face the general downward tendency of rates which, according to some people, threatens to play havoc with their revenue.

This tendency arises in the main from two causes: in the first place the public demands lower rates, in the second place the railways themselves are constantly endeavouring to undersell each other. It will be noted that the two causes to which I assign the movement seem contradictory; I therefore hasten to add that they are not at work side by side, but separately in different sections of the country. In the newer territories the people clamour for lower rates; in the older regions the railways reduce them of their own accord; in both instances the companies are the sufferers.

Let us first take the case of the newer sections, where railways, of course, are youngest: that is to say, passing through the stage which Eastern railways were in twelve or fifteen years ago; for in the

West the railroad history of the East is only repeating itself. Near the seaboard most lines have attained maturity; in the newer regions the majority are still, as it were, feeling their way, and are not yet firmly established. They are, in fact, in the same position as the people amongst whom they exist. The country across the Mississippi, its people and its institutions are at this day still separated from maturity by a wide chasm of time, and that chasm will have to be filled with experience. The object lessons which, as we know, have not been spared to the East are still in store for the Westerner, and notably with regard to railways he has to learn a good many things yet. When he wanted railroads, he was very kind to them, too kind in fact. Now he has got them he is the reverse. He has seen that railways, much though they have accomplished, did not at once introduce those Utopian conditions which he confidently expected, and his disappointment, though natural, has made him cross and unreasonable.

This frame of mind finds expression in some very curious demands. The Westerner—and one might say the American in general—has, in railroad matters, gradually obtained the protection of the law he has so long clamoured for. He is tolerably well guarded against the abuses which in bygone days provoked so much well-founded indignation. But with that he does not seem satisfied. What he requires now is low rates—excessively low rates. In spite of the pamphlets of the boomer—in spite of the grandiloquent leaders of the American newspaper—agriculture does not pay in the United States, that is to say, it does not give the farmer much beyond a fair livelihood, and it does not enable him to indulge in those luxuries which, according to him, surround the Easterners and the people in big towns. It is quite true that he is providing Europe and the large cities of the East with wheat, and that he is pushing the farmers of distant countries out of their own home markets. But he has to overcome great obstacles to do that. In order to compete with others he has to send his cereals a thousand miles and more; and the cost of transportation is so considerable that it reduces the price he obtains for his produce to a very low level, and his profits to very modest dimensions.

Of course, even a farmer's mind perceives that, were there no distance, the cornfields of Kansas and the wheat-lands of the North West would pay better than they do now; and he therefore argues that it is the duty of railways to obliterate it. The reasoning is not very logical, but that matters little as far as the farmer is concerned. He does not care a cent about logic or anything else as long as he gets low rates. He thinks the railway has been built to carry his cereals for next to nothing, and not to yield returns to the people who put their money into it. It is waste of energy to tell him that railway capital is entitled to interest, or that it may look for protection to the people to whom it is entrusted. That does not concern him in the least. Let the capitalist starve, but give the farmer a low rate to the seaboard, no matter whether railroad companies go bankrupt or not.

That is what the process of reasoning of most Western folks amounts to. And the farmer being as yet "boss" of the West, it is not at all surprising that he proceeds to put rates down. He has his Railroad Commissioners, and what on earth is the object of having railroad commissioners if they don't reduce railway rates? So the commissioners do as the farmer bids them. They are drawn from the farming contingent, these commissioners, and sympathise with the movement; but even if they did not believe in its good purpose, they would obey Mr. Hayseed, who appoints and pays them. They know very little of railway business, and still less of railway economics. They are short-sighted, narrow-minded, and believers in heroic remedies. In consequence they reduce the maximum rates already fixed by themselves. They have done it recently in Texas and Nebraska; they did it before that in Iowa, and earlier still in Wisconsin and Minnesota. They know perfectly well that these States at one time paid dearly for such action, so dearly, in fact, that they cancelled their low tariffs with neatness and despatch. But they think the people of Iowa, Minnesota and Wisconsin are "chumps," and honestly believe that they themselves are cleverer and their State better, and so they pay no heed to the object lesson. Of course, after a time they will see the folly of frightening away capital by preventing it from receiving its due, and then they will revise their rates, just as Iowa did. But if you predict this, they laugh at you. And when Nebraska has come to its senses, Wyoming will be where Nebraska is now, and will try the same dodge over again, with precisely the same result.*

This meddling with railroad rates, in fact, is part of the economic education of the Yankee, who never seems to get any sense except by unpleasant object lessons, and never sees the folly of his economic fallacies save after injurious experiments. It is so with McKinleyism, it is so with the Sherman Act, it is the same with railroad legislation. Protection and the silver craze seem to have had their day, and so has railroad legislation, or rather rate regulation, as far as the East is concerned; in the West its day has not yet come, but it is undoubtedly dawning even there in many of the older States. On this side of the Missouri people are rapidly losing faith in rates fixed by commissioners, rates which either do no good or else harm. On the other side of that river people will no doubt also come to their senses, and it will not take them many years either; but they are not quite so far yet. Nevertheless even there this experimenting with tariffs fixed by commissioners who know not what they are doing will not last long. It will soon cease, or rather shift. For, this tendency to meddle with rates is going in a slow wave across the

---

* Since the above was written, the Nebraska Maximum Law has already been suspended.

States, moving westward from the older parts to the newer regions. It has already passed over the East, where it is almost forgotten by this time. It is now at its height about the hundredth degree West of Greenwich. But from there too it will recede. We must expect to see it next in Wyoming and Colorado. Then we shall hear of it in Utah and Idaho. Finally it will sweep over Oregon, Washington, and California, and sink into the Pacific Ocean and oblivion.

But meantime the craze is very unpleasant to the railways. They can if they like carry cases before any of the Upper Courts, and as a rule they will get justice there. At any rate, they won their case against the State of Texas in 1892, the same as they had won similar cases in other States before.* But the abrogation of statutes does not altogether remove the evils of the prejudice. As long as the attitude of the people remains hostile to railroad capital conditions must be strained, and railroad business must be interfered with, especially in America, where the railroad manager is a born fighter. If people make it warm for him, he will insist upon making it warm for the people. He usually gives tit for tat, no matter whether it increases

---

* In August, 1892, a decision was given in the U.S. Circuit Court in Texas, according to which the establishment of obligatory rates by State Railway Commissioners is not a due process of law within the scope and meaning of the Constitution of the United States, and that rates made by States or Rate Commissioners must be reasonable. This decision was given in a test suit brought by the Mercantile Trust Co., of New York, against the State Railroad Commissioners of Texas. The judgment given was that the law fixing the unreasonable tariffs in question is invalid.

the mutual bitterness or not, and without having much regard for the interests of the investors. Strained relations between the population and the railroads, instead of the amicable understanding essential to the welfare of both, are the result, and the hostility affects business in general and railway revenue in particular in a thousand small ways, and makes capital not only unremunerative for the time being but also timid. This deplorable state of affairs will continue until experience brings the people back to their senses; when they have regained these they usually leave railroads alone. Meantime the investor suffers loss and capital grows timid.

With regard to this State interference with rates, and the clamour for low tariffs on the part of the public, it is necessary to make brief reference to one or two other aspects of the question. I find in a report of Convention of State Railroad Commissioners held at Washington in April, 1893, the following remarks, offered by Commissioner Becker of Minnesota, which are admirably suited to serve as a starting point. During the discussion it was suggested that rates should be such as to give a fair return upon the bona-fide cost of a road. I should, by the way, like to see the Commissioner able to fix a tariff, giving the cost of transportation for six classes of freight between every one of a thousand stations and the 999 others, in such manner that it would always yield this revenue to a nicety, despite the enormous fluctuation in the volume of freight;

but that is a detail. Mr. Becker, then, said:—
"Money invested in railroads is no more entitled
"to protection than the private money of any in-
"dividual. I object to an inquiry on the part of any
"commission in this country as to what a railroad
"costs. I do not care what it costs. It may have
"been economically built; it may have been extrava-
"gantly built. I would never inquire how many bonds
"a railroad company had issued. They may issue all
"the bonds they can sell, for all I care. That makes
"no difference. The men who invest their money in
"those bonds, if it is a good investment, will get the
"benefit of it, and if it is a poor one they will lose their
"money. The same with reference to capital stock.
"I should not inquire what amount per mile a railroad
"had invested in capital stock. They may issue all
"the stock they please, and water it all they please, so
"long as they find men to buy it. It is their look out
"whether they have bought good stock and paying
"stock, or whether they have bought poor stock. And
"I will go still further and say that the expense which
"the railroad company is at to carry the freight or
"to manage its business affairs is a matter which does
"not enter at all into the question of what is a reason-
"able rate. *I object most strenuously, as a citizen,*
"*against any form of authority which undertakes to say*
"*that the people of this country are bound for ever to pay*
"*the interest upon the bonds of any railroad company by*
"*freight tariffs, or the dividends upon any stock by*
"*freight tariffs. I think that in looking at the question*

"*of reasonable rates we should not look at it from that standpoint, but we should look at it from the standpoint of a man who pays the freight and see whether his business will justify it to be made.*"

The passage which I have reproduced in italics affords a typical illustration of the mode of thought of your average Western railroad commissioner; but it is so nonsensical, and the procedure it implies would be so monstrous, that it hardly deserves to be taken into serious consideration. It shows that the people, as I said before, do not care a straw for the investor as long as they get a cheap rate—until experience has taught them better. They are too short-sighted to perceive that if they frighten away capital they will be worse off than ever. But we may be sure that they will never succeed in practically confiscating railroad property. Laws as the farmers would have them would not only be ruinous to the entire country because of the withdrawal of every penny of foreign capital, which would be sure to follow, but they would run counter both to the Constitution and to the sense of justice, not to say the common sense, of the American people.

But in the first part of Mr. Becker's utterances there lies a good deal of common sense. In America, as we have seen, the railroad is really a private enterprise, and the Government has no more obligation towards the companies' than towards other property. It must, and we may be sure it will, protect it as such. But it has no responsibility in the matter of returns.

Provided the Government guards their rights of property, investors in railways have only themselves to blame if they make an unremunerative investment. They may build a road which is not wanted, or they may build it badly. They may buy securities at too high a price. They may invest money in a property which lacks honest management. They may commit a dozen other follies. But should one of these mistakes lead to loss the investor cannot reasonably blame anybody but himself. It is the investor's business to see to it that his money is put into a good property, managed by honest people. If he is imprudent enough to neglect that, he himself is responsible for all consequences.

We make these reflections to pave the way for one or two observations related to yet another side of the rate question. They amount to this—that close kinship exists between the popular demand for lower rates and the after-effects of the abuses committed with regard to capitalisation. Everybody knows that American railway capital contains that fictitious element to the discussion of which we have devoted the third chapter. A good deal of the stock issued by the various companies does not all represent money actually invested; and although the American people do not on the whole object to pay such rates as will secure a reasonable return upon the *bonâ fide* cash investment, they are unwilling to submit to charges which will enable the companies to distribute dividends on heavily watered stock. In other words, the people decline to pay the piper, to be responsible for the rascalities perpetrated by the old bosses, for the lack of judgment which resulted in waste of capital, and the like. And these considerations explain utterances like that of Mr. Becker, quoted above.

For, it cannot in fairness be denied that there is some justification for these views. The foremost duty of a railway manager is to earn good dividends upon the capital of his company; and you may argue in whichever way you please, you cannot gainsay that the consideration of a minimum sum, sufficient to pay interest and admitting of the distribution of dividends, must of necessity be taken into account when tariffs are fixed. This means that wherever the capital is inflated there will exist a desire to charge rates high enough to leave returns upon the watered stock; and there exists, therefore, an undeniable connection between water and rates. It does not matter that rates are already lower in America than anywhere else, nor does it say much that to double the dividend it is not necessary to double the rates, but that a fractional rise will do. The fact remains that there is watered stock upon which the roads want to earn returns, and on the strength of this truth we are entitled to maintain that the abuses of capitalisation are intimately connected with, and afford some excuse for, the rate agitation. But the validity of the excuse is apparent only. This water was little else than a premium which the country paid to the promoters and the investors who gave them railways. Had there been no chance of creating water, nobody would have taken the trouble to build roads in a new country, or to advance the money wherewith to build them. Thirty years ago a five or six per cent. American railway bond offered at par would have attracted nobody, and the companies therefore had to choose between watering their stock or paying a high rate—say eight, ten or twelve per cent. To do the last would saddle the young corporations with a weight of obligations which would crush them to death before they reached their teens; to water was therefore the only course open.

Moreover, exaggerated notions concerning the influence of water upon rates are prevalent. There being few companies without it, water is, of course, not altogether devoid of consequences. But its bearing upon rates is not nearly as great as some people seem to suppose. Only a very few roads succeed in earning dividends on watered stock. The New York Central is one of them, the Southern Pacific is another. But similar companies are usually in an exceptional position, and notably the case of the former deserves consideration on the part of railroad commissioners. It is well known that Vanderbilt watered the stock of the New York Central. But this corporation subsequently sunk vast sums in betterments, and paid them out of earnings; the property itself appreciated, and despite all its water, the road is to-day worth fully what it is capitalised at. And there is many a company which is in precisely the same position.

The St. Paul and the Great Northern, for example, belong to this class, and many others; the Northern and Union Pacific will belong to it when their lands and their property have risen in value, as no doubt in course of time they will. Now, will these railroad commissioners contend that such properties are not entitled to reap the benefit of this appreciation? Surely, against Mr. Becker's "money invested in railroads is no more entitled to protection than the private money of any individual" the railroads may set: "money invested in railroads is as much entitled to the benefits accruing from enterprise or sound judgment as the money invested in any other business."

Furthermore, Mr. Becker may safely let the "water *versus* rates" question take care of itself. There is in the United States such an amount of competition that it is impossible for any road with heavily watered stock* to pay dividends upon it. Take, as an example, the competitors of the New York Central, which itself earns good dividends because its water has been evaporated by a rise in the value of the property, and by betterments paid for out of earnings. Amongst its rivals we find the Lehigh Valley, the Lackawanna, and the Erie roads. The Lehigh Valley has a capitalisation of about $98,000 per mile, the Lackawanna of $170,000; both companies have little water in their capital. The Erie has an

---

* By this term we mean, of course, stock in excess of the *actual value* of a road, not of its cost.

enormous amount, its property being capitalised at some $210,000 per mile, though worth much less than that of the two other roads. Now, the Lehigh Valley and Lackawanna earn good dividends, the Erie does not, and it never will unless its water is eliminated by appreciation of its property, or by heavy expenditure on betterments out of earnings. There are several roads situated like the Erie, the Wabash and Reading for instance ; but none of them pays returns upon its watered stock. As a fact, I do not believe it will ever be possible for any road to pay good dividends on watered stock until the water is eliminated by such processes as we have indicated above, and for the following reason : whenever a road earns good returns upon an excessive capitalisation, there exists a strong temptation to construct a competing line, because with less capital, *i.e.*, if built without watered stock, this road would be sure to pay excellent returns. Thus, in theory, profits (and also rates) must be automatically levelled down to the point at which they just pay fair returns upon *bonâ fide* investment; and in practice they are. There do not exist many roads in the United States—no, not even the Erie or the Reading, or the Wabash—which do not pay fair returns upon the money represented by their actual value ; in fact, I do not know of a single one, unless it be still in its infancy. The subtle economic influences that are at work somehow reduce profits to that level, and at the same time they tend to automatically keep rates on a basis which is fair to all concerned, that is, neither so low as to withhold adequate returns from capital, nor so high as to be burdensome to trade and travel. Forsooth, well may Mr. Becker and those whose thoughts he interprets say that the actual cost of a road has nothing to do with the rate it may fairly charge. It is the actual *value* of the property which regulates both its returns and the rates which determine these returns.

But enough of this. What we have discussed here are conditions chiefly prevailing in the West, and perhaps also in the "new" South; conditions which soon, and in the natural course of events, will make room for others. Let us now see how matters lie in the East, where conditions are more mature. Here, too, railroad people have their rate question; but it is of a different kind from what it is "out West." In the Eastern and Central States, and perhaps also in some of the older parts of the North-West, rates are much lower than elsewhere.* Yet they seem still on the downward grade. But this is not a result of rate-reducing railroad commissioners. The railways themselves are responsible for it.

The matter is not at all complicated. The low rates in the East, notably on the Trunk lines, arise almost entirely from competition. It is well known that an amazing number of lines cater for a traffic which, stupendous though it be, does only in prosperous

---

* In 1891 the average freight rate in the Eastern States was 0.78 cents., or less than four-tenths of a penny per ton-mile.

years suffice to keep all lines busy. As a result the roads are as a rule underselling each other, and the consequence is that the rates keep on declining, and that increased traffic often results in decreased net earnings. The Pennsylvania Railroad, for example, the greatest road in America, carried in 1892 13,457,000,000 tons of freight one mile against 12,258,000,000 tons one mile in 1891, an increase of about ten per cent. Owing to the decline in rates, however, the earnings from this traffic increased only from $134,250,000 to $139,000,000, or barely 3½ per cent; expenses, therefore, rose in proportion; and hence net earnings actually amounted to $1,750,000 *less* than in the preceding year, in spite of an enormous *increase* in traffic, and notwithstanding the prevalence of general prosperity!

Small wonder that President Roberts expresses his dismay at this state of things in the observations which he inserts in his report for 1892. After calling attention to the anomalous conditions which cause railroads to earn less with a larger traffic, the President of the Pennsylvania Railroad Co. says:

" In an effort to remedy this unnatural condition
" of affairs, the chief officers of the principal railways
" of the country, after a very full discussion of the
" subject, and at the suggestion of the Inter-State
" Commerce Commission, deemed it proper to appear
" before Congress and suggest such amendments to the
" Inter-State Commerce law as would, while increasing
" the efficiency of the Commission in detecting and

"punishing violations of its provisions, also enable the
"railways to enter into proper traffic relations with
"each other, on a basis to be supervised and approved
"by the Commission. These matters were forcibly
"presented before Committees of both Houses of Con-
"gress, and it was fully explained that the object
"sought was not an increase of rates or the prevention
"of competition, but simply to secure alike to all
"shippers the rates filed with the Commission; but
"the effort to obtain any satisfactory modification
"thereof was unsuccessful. As a result, the greatest
"industrial interest of the country, and the one with
"which its prosperity is the most intimately connected,
"is apparently left in such a position that it is unable
"to enter into any legal arrangements that will enable
"it to meet the anomalous conditions already referred
"to, or to so manage its affairs as to either properly
"serve the public, or make a fair return to its owners.
"It is to be hoped that a careful consideration of the
"subject will convince Congress that the protection of
"the public, no less than of the companies themselves,
"requires at their hands legislation that will authorise
"the making of such contracts under proper supervi-
"sion."

These observations clearly show that the railroads consider themselves aggrieved; but, nevertheless, we find ourselves unable to sympathise with the Pennsylvania president. In the interests of railway investors it is undoubtedly desirable that railways should pay well, but I fail to see that the roads are taking the proper steps to attain this end. It is well known that the trunk line presidents went to Washington chiefly to obtain a repeal of the pooling clause of the Inter-State Commerce Act. But although I have followed their arguments pretty closely, I cannot say I have found them very convincing. It has been abundantly demonstrated that "pools" are neither beneficial to the public nor to the railroads, and the effects of these combinations are so evident that it requires strong arguments to bring the public to the opposite view. And since the American pool is undeniably a bad thing, it was impossible even for men like Mr. Roberts or Mr. Depew to carry the day in Washington. Nor can we say we regret this, since we firmly believe that however undesirable rigid legislative interference on the whole may be—notably State interference with rates—the abolition of pools deserves the hearty approval of railway shareholders.

In the foregoing extract Mr. Roberts says that the object sought by pools was not an increase in rates. But if that is not the case, what earthly purpose does a pool serve? It is admitted that a pool will increase earnings. It is certain that a pool does not increase the volume of freight. And I should like to know how under these circumstances earnings can be increased otherwise than by raising the rates. There is, it will be seen, such an evident hollowness in Mr. Roberts' arguments that nobody can be surprised at his failure.

The views I have just expressed are, it is true, dis-

carded as fallacious by several authorities, whose
greatness I can only measure by my littleness. Mr.
Roberts, Mr. Depew, Mr. Walker, in fact almost every
railroad man, denounces the prohibition of pooling,
defends the practice in general, and recommends its
re-introduction in particular. With such authorities
arrayed against me, I might despair of convincing
anybody of the unsoundness of their views, were it not
that, being railway presidents, these gentlemen are
either prejudiced to an extent which would prevent
them from seeing the weakness of their own arguments, or else fully aware of the defects of their
reasoning, but conscious of the fact that it would be
bad policy to say so. But pools are no novelty in
America; and if they had any merits at all, the presidents who clamour for their reinstatement would not
fail to point them out. Yet I challenge anybody,
first, to prove that pools have ever benefited American
railways, and second, to gainsay that every pool was
injurious to the public. And if pools are of no use to
the companies, and detrimental to the community at
large, why on earth should they be reinstated?

It is worth while to enter very closely into this
question of pools. The presidents are scheming in
Washington to get them back, and will perhaps
be successful. We are told by railroad people that
these pools would be of immense benefit to all concerned, and it may therefore be assumed that investors believe that the repeal of the pooling clause
means the advent of the railroad millenium. Yet

it may be regarded as certain that this expectation is not warranted, and that pooling would be quite as disastrous to the investor in the future as it has been in the past. That is why we think we must go fully into the matter.

Let us see how these pools work. We will take a trunk line pool. There are four direct routes between Chicago and New York; there are nineteen indirect lines. The Vanderbilt, the Pennsylvania, the Baltimore and Ohio, and the Erie roads are the best. Given equal rates without a pool, and no sane man will send anything or travel by any of the roundabout routes. The bad lines know this very well, and therefore they lower their rates to such extent as will counterbalance the drawback of their inferior service. This is done by consent of the good lines, who deem it wise to make this concession for the sake of peace. Thus arises a condition which is fair both to the public and to the roads. The public can get all kinds of transportation: it can spend $30 on a first-class ticket from New York to Chicago if it wants to travel by vestibuled express, and it can get there for half that sum if it is satisfied with a steamer to Portland, Me., and with the roundabout service of the Grand Trunk Railway; you can send a case of woollens, if wanted to be delivered within thirty hours, by the N.Y.C. at N.Y.C. rates, but if time is no object, you can let it go by steamer to Norfolk, and then by the Cumberland Gap express, over the Norfolk and Western, the Louisville and Nashville, and the

Evansville lines. You can also strike the happy medium provided by the Erie, the Lackawanna-Grand Trunk-Wabash combination, and others. This is fair, and as it should be; you pay according to the quality of the service, and the roads are remunerated on the same just basis. But trade seeks the best routes, and the roundabout lines do not get much traffic, in spite of their lower rates. They want, however, more of it, and therefore start cutting. True, they lose over the transaction, but so do the better lines; and in consequence a pool is attempted. All lines make the same rate between two points, and apportion the traffic. The result is that the public has to pay wholly or nearly the same price for good and bad transportation alike, and that it cannot say whether the goods it sends shall go by a fast $1\frac{1}{2}$ day line, or by one which may take three weeks, as some do. Furthermore, all lines have their fixed proportion of the traffic, and can get neither more if they exert themselves, nor less if they provide the most inferior service. As a result the latter becomes worse the longer the pool lasts. Now we need not even remember that pools cause severe fluctuations in rates because they are shortlived and lead to rate wars which upset trade; what precedes is quite sufficient to demonstrate that pools are bad from the public's point of view, and that the universal dislike with which they are regarded is very well founded.

Now let us look at the matter from the railways' point of view. Without a pool—*i.e.*, when free competition determines the cost and quality of transportation—the good line, with a small capital and a shrewd management, is best off, as it should be; I say as it should be, because every man and every company is entitled to the benefits accruing from foresight, economy, good management, and the like. The weak road may succumb—usually *does* succumb. But does this matter? Is it the duty of the public to support a shop which is badly managed, lays in an unsaleable stock of goods, has a heavy mortgage, etc., in order that it may keep out of the Bankruptcy Court? Should a nation be forced to keep alive a road which was built without being wanted, robbed by the adventurers who built it, and is doomed to ruin if matters are allowed to run their free course? If such a road wants to cut rates, let it do so by all means : if only the other roads make a firm stand it will do little harm, and kill itself all the sooner. It will go bankrupt, or be reorganised, or perchance be bought by one of the better roads, which will keep it for its cheap and slow traffic, like the New York Central does the West Shore. "But this would amount to reckless waste of capital," cries your "railway economist." It would not, and even if it did it would be deliberate waste on the part of those who called the road into existence, waste which was foreseeable, waste for which the public cannot be held responsible, and waste from which the public will not suffer. I will admit it is hard for the investor who loses over the affair; but it is his look out that he does not get interested in roads, bad, superfluous, or dependent upon dishonest directors. If he does not take care he cannot blame others if he loses, nor expect others to guard him against damage. It would be nothing short of monstrous to compel the public to pay the penalty for the folly of the investor or for the rascalities of directors. Yet even the good roads will try to keep the bad ones going. They form a pool. They themselves do not benefit by it; they only pay blackmail to a weak road which might temporarily reduce their business. They do not, or will not, see that this momentary loss would be ultimate gain. They form a pool to keep up their customary dividend without a hitch, and to maintain the price of their stock at the highest possible level. Of course, the pool does not last long. Every one of its members *can* break it up; several of them *will* break it up as soon as it suits their book, for instance, when they have "beared" a sufficient amount of their stock in Wall Street. To make their profits in Wall Street all the bigger, they will cut to a phenomenal extent. The cut will be "met" by others, "to drive the rascal back into the pool." And one week of universal cutting will eliminate whatever increase in the revenue may have been brought about by the pool. That was the experience gained in the days of pools. Yet—*O tempora, O mores!*—railroad people ask for the repeal of the "fifth clause"—Mr. Depew and Mr. Roberts as well as the small fry of presidents and managers.

It will be asked how I reconcile this fact with my opinion that every railroad man knows that pools are of no use to his company; and I will state what I think is the reason. I ascribe it partly to personal motives of the American railroad president, partly to his exaggerated conception of the virtues of his colleagues. The lower class of presidents yearn for the return of the good old days of pools, when they could make money in Wall Street. The better class yearn for the restoration of their vanishing powers, and in addition seems to think that, pools being applied with fair success in other countries, they must be equally successful in America. But this assumption is a mistake. A pool is a conspiracy of railroads, having for its object to force a nation to pay excellent returns upon the capital employed by its railways. Where there is a highly developed business morality and a jealous legislative supervision, as in England, they may be fairly successful and comparatively harmless. With a young impulsive nation like the Americans, whose business morality has not yet reached its highest attainable point, they must end in disaster. I think the American presidents ought to see that; at any rate I cannot perceive how any one of them can for a moment believe that a single company will fare the better for pools. It is impossible to give the name of only one company which saw its revenue increase in proportion to its tonnage during the era of pools, for so brief a span as twelve months.

What strikes one most with regard to pools is the insane desire of the roundabout routes to get what they call their share of the through traffic; for pools only regulate through traffic, and do not touch local business. It is evident that where the New York Central can only get a scanty profit with its low rates and excellent lines, the Grand Trunk cannot make a profit with rates still lower, with a much longer route, and a very inferior roadbed. Yet all roads want to get "through traffic" to which their situation does not entitle them. When will they adopt the wiser course and concentrate all their energies exclusively upon the lifeblood of railways—local traffic? This fierce fight for through business has brought dozens of prosperous companies to rack and ruin. The whole list, from Atchison down to Wabash, bears the sad traces of this stupid struggle—this fruitless fight which is one of the greatest evils of American railways. And the worst of it that in this respect companies of the highest standing are almost as bad as those of the lowest; the Pennsylvania, for example, makes no exception, as will be seen from the following paragraph which we are surprised to find in the last Pennsylvania report immediately before the grievance paragraph already quoted on p. 73:—
"The policy referred to in the last annual report, of "stimulating, through an increase of equipment and "other facilities, the movement of grain between Erie "and Philadelphia, resulted in a largely increased "volume of traffic; but owing to the sharp competi-

"tion in the rates between the lines engaged in lake
"and rail transportation, these efforts would seem
"to have been more beneficial to the commercial
"interests of the city of Philadelphia than to those of
"the transportation companies."

The question which forces itself upon us is, Why does the Pennsylvania cater at all for this traffic? The president frankly avows it does not pay, and he ought to have known this beforehand, in 1891, before he spent money on "the increase of facilities," as well as in 1892, when he had to acknowledge that the expenditure had been wasted. Nobody can be in the least surprised at this failure. Grain from Erie to the seaboard ought to be left to go along its natural route, that is *via* Buffalo, the Lake and the Erie Canal; if in spite of the low rates at which it must be carried, the Pennsylvania endeavours to force it over its own road to Philadelphia, across the mountains, and loses over the transaction, it cannot reasonably complain.

It is this foolish competition for unpaying through freight which is the ruin of American rates. No railway has any inducement to compete for such business. The companies should foster local traffic. If they do that they will obtain good rates which pay, and though their traffic may be smaller their profits will be larger. The New England roads, the oldest in the country, have a strong local traffic, and care little or nothing about through freights; they get an average rate nearly four times as large as the other Eastern roads, and they pay large dividends in consequence. Why cannot the other systems follow their example? Surely they must know by this time that local traffic is the lifeblood of a railway, and that the game played for through business is not worth the candle. There is not a prosperous road in the country which has not a strong local business, and no line living on through freight is doing well. Yet, whenever and wherever railway rates have come to grief, it has been because the railways were cutting each other's throats in order to obtain unremunerative through traffic. Why not put a stop to this foolish policy; why not devote all energies to local business? It is all very well to abuse railway commissioners. But the harm these gentry have done to rates is nothing compared with the injury which railways have inflicted upon themselves by their fierce and foolish fights for through traffic.

The railways know this full well; and those who follow the development of traffic matters in the United States must be glad to observe that the painful lesson which three decades have taught is gradually beginning to bear fruit. Rate wars in their fiercest forms —and these wars naturally were always fought for through traffic—are abandoned now, and in the contest for through business the lines are one after the other throwing up the sponge. A few devote themselves almost entirely to local freight; others do so chiefly, but keep a finger in the through traffic pie; only three or four lines of low standing try to subsist principally on through business, the refuse of railway traffic.

Thus there is improvement in the rate situation, as well as in other matters; the way is being paved for sound rate conditions, and a few companies have already reached this goal. There is not a single feature in the entire complex question of rates which need cause alarm. The people may for a time meddle with rates and give directors and investors a bad quarter of an hour; the companies may act foolishly; but experience will ultimately bring both to their senses. For this experience the investor may have to pay occasionally whenever he ventures outside the securities not exposed to the vicissitudes of the rate situation. But he need not run similar risks unless he feels so disposed; and there can be no doubt that in the long run matters will turn out all right. The roads will not break their own windows by rousing the ire of the people; the people will do nothing which would frighten away capital. And with the corporations careful not to give renewed offence, and the people inspired with a wholesome dread of excessive or unjust interference, the matter may safely be left to settle itself. There is the less reason to feel alarmed because satisfactory conditions prevail in New England, whose railways, now matured, have gone through all the stages younger roads are in yet. There the lines supply an efficient service, get rates four times as high as the trunk lines, and pay heavy dividends; the rigid supervision of the various States does not in the least interfere with the adequacy of rates or the prosperity of the companies. There is no reason why the same thing should not happen in the other sections of the country; only we shall have to wait till they too have attained maturity. During their era of growth we must expect unpleasantness, losses and risks; but it is in return for these that we get higher interest upon our capital. Nor should we overlook the fact that even to-day there is, in spite of the rate question, no line of any importance which does not pay handsome returns upon its *actual cost*.

CHAPTER V.

REVENUE AND ITS APPLICATION.

We have hitherto spoken chiefly of matters affecting railroad income; the uses to which the companies put their revenue is to engage our attention next.

It seems desirable to introduce this subject by referring to one or two of its most salient features. In the first place it may .be well to point out that whereas there is but one way of obtaining revenue— that is, by earning it—there are several means of disposing of it; railways can either spend their income upon betterments, invest it in some way or other, or distribute it amongst the owners or creditors of the property. In the second place it seems expedient to state that the application of revenue largely rests in the discretion of railway managers. Railway directors in the United States are endowed with an extraordinary amount of discretion, so much so that the free hand they enjoy is one of the principal points investors have to reckon with. I have laid stress upon their great discretionary powers by devoting a special chapter to the subject, in the hope of firmly impressing its bearing upon the reader's mind. And I should like to call attention to it once more, since this discretion seems to affect the application of revenue more visibly, more frequently, and more deeply than it affects anything else.

I say seems to affect; for in reality these great discretionary powers have had a far greater bearing upon capitalisation than upon the application of income. Perhaps this does not seem so, unless the matter is gone into rather thoroughly. Nevertheless reflection will soon bear out the statement. There cannot be any doubt that but for the free hand left to directors a goodly proportion of the water now existing—in fact that entire part of it which was avoidable—could never have been poured out. Now this water, and therefore the discretion of directors which was instrumental in creating it, permanently affects the bond and stockholders; they are reminded of it every year, and every balance sheet, every income account, every dividend, shows its influence; it leaves its trace throughout the entire history of a road, and will impress its stamp upon the future; it affects even rates, physical condition, and earning capacity. Its effects are destined to last for ever. But the discretion which disposes of revenue is not permanent, only transitory. It may either exercise or waive its powers at any time, and will leave no permanent track behind, except in a few abnormal cases. For that reason it would be of subordinate importance if it could not and did not assert itself at any moment. The influence of this discretion upon *capital* is therefore only as it were occasional, though far-reaching; its bearing upon *returns* is frequent, but of less stupendous consequence; and hence its results appear to be much more baneful with regard to the application of revenue than with regard to capitalisation, though this appearance is deceptive in the extreme. However, small but repeated inflictions usually seem much greater than single but heavy blows with lasting effects.

We referred above to the well-known fact that ways of spending railway income are three in number: it may be disbursed for investments, betterments, or dividends. Let us consider the effects which unlimited discretion may have upon each. It will be best to illustrate our remarks by referring to the affairs and accounts of an imaginary railway, and we will therefore assume the existence of the A. B. and C. Railroad Co., with a funded debt of $10,000,000, bearing interest at the rate of 5 per cent., and a share capital of an equal amount, divided into $5,000,000 five per cent. non-cumulative preferred shares, and $5,000,000 ordinary stock. We will further suppose that this railway has annual gross earnings to the extent of $3,000,000, whilst its working expenses amount to $1,900,000. With this simple case before us, it will be easy to find lucid illustrations whenever we require them.

Let us first take the competence of directors to invest money. Their road, it will be observed, is quite a paying concern, though not a very large affair in a country where some companies represent as much as $500,000,000 capital. It earns enough

F

to pay its interest, to distribute 5 per cent. amongst its shareholders, and to build up a nice little reserve. Under normal conditions, and with an ideal management, its income account will therefore be something like the following :—

INCOME ACCOUNT OF THE A. B. AND C. RAILROAD.

Gross earnings	$3,000,000
Working expenses	1,900,000
Net earnings	$1,100,000
Interest on bonds ... $500,000	
Dividends on preferred stock ... 250,000	
„ common stock ... 250,000	
Total disbursements	1,000,000
Surplus	$100,000

This is the account under normal conditions. But now the directors get the extension bee in their bonnet, and lease or acquire control of a road. They have perfect authority to do this, and similar investments or acquisitions are made every week, almost every day. Sometimes they are on a grand scale, and sometimes only of small moment; but they can be made, and are made constantly, by directors in all parts of the country who consult nobody but their colleagues.

Now let us suppose the investment or lease turns out well; then, of course, the *ordinary* shareholders of the A. B. and C. reap the benefit; to the preferred shareholders it makes no difference. But leases and investments do not always pay; in fact, a road is never leased to another company unless its directors have a reason for it, and that reason usually is that the lessee guarantees better returns than the road could earn independently; the lessee hopes to make his profit by the economy or by any of the other benefits that usually attend amalgamation. But it happens quite often that the lessee closes a bad bargain. Let us imagine that the A. B. and C. Railroad Company has made such a mistake, in consequence whereof it loses $500,000 per annum. The result is that the common shareholders no longer obtain a dividend; but they would have had all the profit had the lease or the investment been remunerative, and at any rate had a chance of gaining. But as a further consequence of the blunder the preferred stockholders only get $100,000 amongst them, or 2 per cent. on their shares, instead of their accustomed 5 per cent.; and on these people the loss is rather hard, for their dividend being limited to 5 per cent. they run the risk of loss through the lease without any corresponding possibility of benefit. Matters may even get worse than this. Next year the A. B. and C. may lose $700,000 instead of $500,000; and then it will be unable to pay the interest on its bonds, and must go bankrupt unless it happens to have a reserve to fall back upon, or takes to borrowing. All that will be the result of the unlimited discretion of directors.

Let it not be supposed that this case is an exaggerated example. Similar occurrences are quite common in American railroad history, and dozens of companies either have suffered through such leases or still sigh under their burden. The Louisville and Nashville has twice been forced to reduce its dividends on account of leases and extensions; the Atchison came to grief mainly through extension; the Reading fell three times because of unremunerative investments which, to aggravate matters, seem to have been dishonest investments into the bargain. The Norfolk and Western had to stop its dividend in 1892 chiefly because it went in for extension, and there are a score of other equally prominent companies like the Erie, the Wabash, the Denver and Rio Grande, which either have suffered or still suffer from the effects of similar unfortunate ventures.

Against these abuses the investor cannot protect himself. He must trust to the integrity and judgment of his directors; where honesty and ability are wanting he is likely to become a helpless victim, unable to do anything except bewail his losses. And even where these two desirable qualities are not lacking he may suffer. His directors may be honest and able, and yet they may incur risks and land their employers dry. They may embark upon new enterprises, notably extensions, in the firm belief that they are furthering the best interests of those whom they represent, and nevertheless may plunge their property into ruin. The same may of course occur in cases where shareholders have greater control over their property than in America with its peculiarly autocratic business management; for after all shareholders are guided by, and as a rule rely upon, the advice of their directors. But wherever a company's officers need not consult their shareholders every time before they decide grave and vital matters, there is apt to come into existence a stronger tendency to venture than in those cases where stockholders do something more than elect directors who during their term of office can do as they please. The consultation of shareholders by directors may be what it is sometimes called—a mere sham; but nevertheless the mere formality seems to increase the consciousness of responsibility and to cultivate caution in directors. In America directors have *plein pouvoir*, a mandate without instructions; and though they are in a way responsible, the fact that at the worst they can only be censured after they have blundered seems to blunt their sense of responsibility and to foster their innate tendency to embark upon all kinds of extensive enterprises. Most Americans feel pretty big anyhow, especially when they happen to be railroad presidents. As a class the latter are bold and ambitious, and not being restrained by cumbersome instructions, by vigilant supervision, or by a keen sense of responsibility, they would embark upon big schemes even if they were prompted by nothing but professional ambition, although that is nowise the case, as we shall see presently.

The shareholder is less ambitious in his ideas; in

fact, it may be safely assumed that he does not care for these constant extensions in some form or other, of which his directors are so passionately fond. Investors as a rule are of a cautious temperament, though they do not always act cautiously. And to confine ourselves to the example of our imaginary A. B. and C. Railroad Company, those financially interested in it, though not adverse to receiving higher returns upon their capital, or to an increase of the safety of their returns, must be supposed to prefer their regular 5 per cent. to the chance of getting more which they may have in return for the risk of losing all. They are apt to ask: Why not leave well alone? And were they consulted they would, unless they were fools, say that they are satisfied with the safe and reasonable return they enjoy, and instruct their directors to confine their energies exclusively to endeavours tending to increase these returns by a conservative improvement of the property. To your average shareholder good and safe returns are the main point. He is not sentimental, and he does not care a pin whether he draws his 5 per cent. from an obscure company or from a railway "chaining the Pacific Ocean to the Great Lakes," or performing some other grand feat.

Not so with the majority of managers. To the railroad director it matters a great deal whether his company is large or small. A road only a few hundred miles long cannot afford to pay the princely salaries with which the great systems endow their first officers; and, what says still more, the president of a small company does not have that great and gratifying power which a Chauncey Depew or a President Roberts has, and which is so dear to the heart of the American. And the knowledge that the managers of big systems receive big salaries, and rank among the most influential men of the Republic, has a great deal to do with the development of vast systems, and with the constant amalgamation and extension of smaller ones. Look at it from whichever point you may, you will be unable to find a more potent cause of this constant sacrifice of prosperity to greatness, which is so conspicuous a feature of the American railroad, than the personal ambition of the presidents.

Nevertheless the tendency is often defended, and defended with a good deal of ability. If you make observations like the foregoing to an official who has caught the extension fever, he will overwhelm you with a variety of apparently valid and convincing arguments in favour of it. He will tell you that unless his company builds branches into the adjacent country, rival systems will construct them, and that, in consequence, his system will be a tree without roots when the country has grown up. There is something in this, but only a little, for the best systems have not many branches; they are either lines with few feeders but a fine strategic position, or roads which prosper chiefly because they command all the business of a certain district; your "through route" of a thousand miles or so never pays, as we have seen in the preceding chapter. Then the manager will say that the wily rival may "obtain control" of your road, and either play ducks and drakes with your dividend, or kill you by means of a rate war. There is some truth in this too, but only because the American system of "bossdom" with its incessant changes of ownership and "control," and with its end-justifies-the-means policy, is a very bad one. Then the circumstances calling for and leading to consolidation are often referred to as an excuse for this everlasting sacrifice of prosperity to the Moloch of greatness and greed. But it is all rubbish. There can be no excuse for a policy which ruins prosperous properties by the dozen simply because they are small; not even the chance of acquiring a prosperous greatness can justify it.

As we have said, the worst of all this is that the shareholder has no say in the matter. If a board has ruined a property, the shareholder may refuse to re-elect it at the next annual meeting, but after all it is poor comfort to shut the stable door after the horse has bolted. The shareholder ought to be guarded against these "shocking outrages." But how is he to be protected against them? Even if he were consulted before these decisive steps are taken the recommendation of his directors whom he trusts*— for if he did not trust them they would not be his

---

* In theory, of course; and it seems to go without saying that we allude to the election of directorates by majorities of shareholders.

directors—might lead him into misfortune. Hence the only thing that can shield him is the honesty and prudence of those entrusted with the management. Fortunately the number of boards possessing these essential qualities increases with the improvement of business morality, and with the gradual though somewhat slow spread of conservatism in business methods. There are by this time quite a number of companies whose boards deserve the most unqualified confidence; twenty years ago there were none. Their number will, I have no doubt, keep on growing; but in the meantime the nuisance continues to exist, and the investor is apt to suffer from it unless he is on his guard.

We now come to the second part of our subject —the discretion directors have over expenditure for improvements. In this respect we have nothing to do with motives like those which are responsible for disastrous extension. There is no room for dishonesty here except in those cases where accounts are manipulated in connection with speculation, and these instances are comparatively rare, though they used to be common enough. Nevertheless the practice is fraught with danger, as we shall show presently. Let the reader turn to the statement of income on page 90; it will facilitate reference. Now let us suppose the A. B. and C. Railroad requires improvements. Traffic has grown, and it is desirable to replace its light and worn-out rails by heavier ones, and a few wooden trestles by iron structures. The expense entailed will be, say, $500,000. How is this expenditure to be arranged for?

Three courses are open to the management. It can borrow the money, and consequently increase its capital or its debt; it can provide for the expenditure out of its reserve, if it has one; it can pay for the betterments out of income, and be forced to reduce its dividends in consequence. Only a few companies have reserve funds admitting of anything except equalisation of dividends, and large enough to pay for heavy betterments. No doubt the gradual creation of special betterment-funds would be desirable, but there are as yet various obstacles which prevent their formation. If a certain sum could be set apart each year to pay for costly improvements when they are needed, there would be no necessity to subject either the capitalisation or the returns upon capital to sudden interference. But the man who puts away something for a rainy day must either have some income to spare or else practice self-denial; and American roads are as yet neither over-prosperous, nor capable of reducing their wants. So we may leave this desideratum out of the question, and then we can choose only between an increase of capital or payment out of current income.

Something can be said in favour of each method, though both have their peculiar drawbacks. But on the whole the issue of fresh capital is, in the abstract, by far the more desirable course in all cases where the betterments amount to the introduction of improvements capable of increasing revenue ; not of course where they are merely repairs. Renewals are part of the working expenses, and inseparable from conducting railroad business; and it would be absurd to pay for them out of capital. But as to improvements which increase earning capacity without being merely repairs, these should be charged invariably to capital account. This method, however, has one disadvantage, though perhaps only in the eyes of those who are wont to look far ahead : it results in a constant increase of capital, and the directorate being of course desirous of keeping that capital productive, will have to charge heavy rates when this method has been in vogue for many years. It has the effect of gradually piling up an enormous capitalisation, a fact abundantly demonstrated by English experience. In this country roads earn considerably more per mile than in America, and their expenses per cent. are much smaller ; but the capital employed in England in proportion to the volume of traffic is stupendous in comparison with America, and as it is the aim of directors to earn dividends upon this vast capital, the constant increase attending the English practice is largely responsible for the exorbitant rates charged in this country, rates which would be impossible in the United States. Were it to charge the average English tariff, the American railroad would be useless, at any rate at present ; and hence it is doubtful whether the policy can safely be followed by American railroads to the same extent as by English companies. For, in fact, it has been adopted to a small degree. Betterments often require such heavy outlays that to pay them in a lump out of earnings would be impossible or inexpedient. It is equally undesirable to issue new capital, and hence floating liabilities are incurred. This floating debt is often funded—usually in fact—and hence betterments do really to some extent increase capitalisation in many cases where they do not seem to. And to mention another aspect of the question, although in theory improvements should be charged to capital account, it is not always possible to do this in practice. What, for example, should a company with an exhausted credit do? Improve its road it must; if its track and rolling stock are not kept in as perfect a state as possible, profitable working will soon be out of the question, and the receiver in possession. And a company similarly situated has no option but to pay for its betterments out of earnings. The Erie's position illustrates this case very forcibly.

In a recent number of a London newspaper[*] I find some very concise and cogent remarks on this complex betterment question; and they are so admirably suited to my purpose that I reproduce them here. The newspaper in question says: " By drawing such funds from revenue, the capital of the company, yielding productive returns, is really being constantly increased,[†] whilst its nominal capital account remains

---

[*] *Financial Times*, June 2nd, 1893.
[†] See the remarks on the creation of water to represent similar expenditure in the capitalisation on Chap. III.

unaltered; and in this respect the American system offers a marked contrast to our own. The custom of English boards is to draw a line clearly between expenditure on capital account and that which properly falls on revenue. All improvements of a productive character capable of earning interest are carried out by the company making a further issue of capital stock, while revenue has to bear expenses requisite to keep the permanent way, rolling stock and equipment in an efficient state of repair. Instances must sometimes arise where the distinction between revenue and capital charges cannot be precisely drawn. If iron rails are replaced by steel, that part of the outlay which represents the replacement of worn-out rails by new ones of the same character, is a charge falling upon revenue; but, inasmuch as the steel rails possess greater durability, their productive yield is greater, and therefore to this extent the cost should fall on capital. After deducting all items properly chargeable to renewals, the earnings of English companies are, almost to their full amount, divided each half-year amongst the stockholders; the directors carrying forward only a relatively small surplus."

" In America, all attempts to preserve this important distinction between revenue and capital charges are, in the great majority of cases, wholly abandoned; and surpluses sufficient to provide a more than fractional dividend are constantly devoted to capital outlay. The larger and more prosperous roads, such as the Pennsylvania and the New York Central, have erred in this respect equally with those companies which are in such poor credit that it is a matter of great difficulty for them to raise fresh capital, and who are, therefore, practically compelled to trench upon their earnings in order to make imperative improvements. If the capital stock of these companies is divided into mortgage bonds with foreclosure rights and common stock only, no serious injustice is done to the latter by depriving them of dividends in order to make additions to their property; for, if such additions were not made, increasing competition might render the line incapable of keeping its head above water, and the common stockholders have, at least, as great an interest in avoiding a receivership as the mortgage bondholders, whilst the reversionary interest in the yield on capital so derived from revenue belongs altogether to the common shares. Such a method of finance results practically in taxing the present for the sake of prospective advantages; but although the common shareholder may cavil at such enforced thrift, and grumble that "while the grass is growing the steed is starving," no grave wrong is inflicted, because even if fresh capital could be raised by the company, it could only be in a form having priority over the common stock. But where the capital stock of the company comprises income bonds or preference shares in addition to mortgage bonds and common stock, this method of making productive improvements out of revenue may act very inequitably indeed. If, for instance, the earnings admit of a dividend on income bonds being paid, and this fund, instead of being so applied, be diverted to "betterments," the result is that the revenue belonging by right exclusively to the holders of one class of security becomes an addition to the capital of the whole concern; and improvements so made may in the future enable a dividend to be paid to the holders of the lowest grade of securities, who have not contributed anything towards the cost of the additions to which the increased dividend-earning power is to be attributed. It is to tax some for the benefit—wholly or partially—of others, and, therefore, constitutes a grave hardship to the holders of income bonds or preference securities."

"These considerations alone make it very desirable that this "betterment" system should be abandoned in all cases where the credit of a company admits of the English method being adopted; but they are not the worst evils that result from the custom of improperly applying revenue to improvement of the road. The necessity for such improvement being entirely in the discretion of the directors, an opportunity is furnished for any amount of jobbery and wire-pulling, which the shareholders are far less competent to check than they would be if they were called upon to sanction a further issue of capital. Directors may one year show the most glowing results by starving the line of necessary renewals, while in the next large gross and net increases in the earnings may be accompanied by diminished dividends, in order to allow the wire-pullers to operate successfully on the Stock Exchange. The discredited position occupied by even the better class of American common stocks in the minds of investors is largely due to this cause, and it is useless to expect that their investment value will appreciably improve until a sounder, or at least a less uncertain system of finance, shall become fashionable with American boards."

Well, that sounder system will no doubt come in due time; in fact, it *is* coming, for recent railroad history shows most conclusively that the foundations of it have been laid. The best companies, for instance the New York Central and the Illinois Central, appear to have altogether abandoned the practice of charging productive betterments to income; both have in recent times issued stock to defray the cost of betterments, and there are several other companies which have done the same, or would do the same if there were occasion. What says more, there can be no doubt whatsoever that the new system will spread. But you must give the companies time. Rome was not built in a day; and the methods of English railway finance were not perfect from the start either. They have been shaped by time and by experience, and in America it will be the same. Meantime we can only say with regard to this matter what we have already said with regard to others— let the investor be cautious. It may be a very commonplace piece of advice, but it is not superfluous. Knowing that defective principles exist it is his business to avoid all stocks likely to suffer from them. Nobody should put money into a stock the dividends on which may be interfered with by "betterments." And those who disregard this maxim must not complain when its soundness is impressed upon them by loss and annoyance.

But there are matters besides betterments which may interfere with dividends; and their existence leads us on the third and last part of this chapter.

A dividend may be earned and ready for distribution, and yet it may not be paid. The directors have absolute discretion over it unless their powers are limited by prescriptions like the usual stipulations of preferred shares; in such cases it must be paid when earned, and if the board wish to dodge this duty they have to fall back upon the "betterment trick." But with ordinary-share dividends directors can do as they please, and if they are bent upon disappointing people there is nothing to prevent them. But this power is not abused very often. Only dishonest directors will put their discretion to such miserable uses, chiefly to further their success in Wall Street, and instances of this despicable practice are happily becoming very rare. At one time they used to be quite frequent. Did not Drew and Vanderbilt make the most of their power in this direction, and has not Gould unexpectedly passed his Missouri

Pacific dividend as recently as 1891 ? However, the Commodore and Gould have had their day; and everybody who has been acquainted with American railroads for a period of any length will be able to testify to the salutary change that has come over the dividend question, unsatisfactory though it may be even at present. Unless the funds regarded as available for dividends are urgently needed elsewhere they are usually distributed, though rarely up to the hilt as in England. In recent years there have been but a few isolated complaints on this score, and each time in the case of a company in which the public anyhow had no confidence. But it has repeatedly happened that a company has for some time deferred the resumption of interrupted dividends, in spite of some surplus available for that purpose being earned. The St. Paul is a notable instance. But these steps were invariably justified, because the object was evidently to secure continuity and regularity of dividends. It is all very well to talk about dividing profits up to the hilt, as is so often done in England. That is right and well in an old settled country like ours, where earnings do not fluctuate very much. But in America, with its frequent, sudden, and severe fluctuations of trade it is a different matter altogether. The annual division of all profits would make dividends still more irregular than they are and cause excessive fluctuations in the prices of stocks; these changes may suit the gambling fraternity, but they can scarcely be approved of by genuine investors.

CONCLUSIONS.

The inferences to be drawn from the vast discretion of managers, which this chapter has illustrated, are plain, and can be summed up in a few words. American railway managers have absolute control over the application of that part of the revenue which is not claimed by protected securities like bonds. They may decrease or destroy their surplus by extensions; they may apply it to betterments; they may abstain from distributing it, even if they cannot put it to other uses; it is only a question of either honesty or judgment, or perhaps of both. Therefore let investors, as much as they can, beware of all those securities which come under the influence of this discretion. As a rule such stocks will consist of common shares, and, in the case of weaker companies, of preferred shares, perhaps also of income bonds, which are only shares with a pretty name. But an exception may safely be made in the case of shares issued by companies of undoubted standing, like the New York Central, Pennsylvania, Illinois Central, Chicago and Alton, and a few others. In these instances there is no risk of loss through dishonesty or bad management. There is only that liability to more or less excessive fluctuations in dividends which are caused by changes in the condition of trade; changes which are always more frequent and severe in young countries like the United States than in the old and stable communities of Europe.

CHAPTER VI.

SECURITIES AND THEIR RETURNS.

We have been careful to state, with as much stress as we possibly could, that the vast amount of discretion vested in directors has most indelibly impressed its stamp upon the capitalisation of American railroads—one might say has framed it altogether. And considering that this discretion is without a parallel it can excite little surprise that the capitalisation was placed upon an altogether exceptional basis.

Experience soon taught the investor what consequences would probably ensue from the virtually unrestricted power directors were endowed with; it may even be said that the tendencies it might engender, and the results it might lead to, were clearly foreseen before any money was put into railways. Knowing himself to be at the mercy of the directors, of his own debtors or servants, the investor exercised caution from the very start; and as time went on he resolutely endeavoured to restrict the powers of the managers as much as circumstances permitted. The managers, on the other hand, resisted these attempts with equal determination; and the result has been that the capitalisation of American railroad companies clearly bears the impression of this struggle for supremacy and power, and became quite as varied as it is peculiar.

The caution of the investor had the well-known result that bonded debt became the financial foundation of almost every one of the hundreds of vast corporations now controlling the transportation business of the United States. The latest Parliamentary return issued computes the proportion of loans to the stock and share capital in England at 26 : 100, and nearly all these loans, which correspond with the bonded debt of American roads, were contracted after the property had been built. In the United States the bonded debt reaches a total of almost £1,100,000,000, and contains little water; but the share capital, though nominally amounting to nearly £1,000,000,000, represents a cash investment of certainly not more than £100,000,000 sterling, and the roads were built almost entirely with the proceeds from the sale of bonds. The contrast between the English and American practice is therefore most striking.

This difference arose chiefly because it would have been impossible for American railroad companies to have raised an adequate share capital. Investors displayed little or no inclination to become owners of the properties, because in that case they would have had to share all risks, apart from being at the mercy of their directors; hence they sternly refused to become anything else than creditors, and, in addition, they would only advance money against security. It need scarcely be said that the promoters had to accede to these wishes, for otherwise they would have obtained no money. Apart from this, it lies in the nature of things that the borrower should yield to the wish of the lender, and that companies should accommodate themselves to the markets their securities are likely to have. Thus the American railroad bond was brought into existence, and became the keystone of the capitalisation of the companies. But the difference between English and American railroad capitalisation does not stop at this point. In this country, the obligation (the debenture, &c.) is secured only by a prior lien on the *revenue* of the company. In America it has first claims, not only to the income but to the entire *property*, and a railway mortgage differs but slightly, and only with regard to immaterial points, from a mortgage on land or houses. A railway mortgage deed is a contract between the company and the investor—or rather the trustee, his representative—violation of which gives the injured party the same claim to redress as the breach of any other contract. If the company does not promptly pay principal and interest when due, the bondholders can foreclose the property; and, if the company should reduce the rate of interest, or modify, or alter, the terms relating to repayment of the principal, the bondholder can seek the protection of the Courts, and will obtain it beyond a doubt.

At this point we are exposed to a strong temptation to deviate into a side-issue of considerable importance. It happens occasionally that a company reduces, or endeavours to reduce, the interest on its bonds, notably when it is being reorganised without foreclosure; and these affairs usually cause so much stir that the reader of this book would no doubt attentively peruse a lengthy dissertation on the subject. Nevertheless, we do not propose to enter into a similar discussion; our space is limited, and it would be next to impossible to deal with the question in all its phases. But we may briefly state one or two general principles which will be found to apply to all cases. It is not necessary for a bondholder to consent to any such reduction, although occasionally, perhaps usually, it is wise to do so. In cases where a road cannot earn the interest on the issue to which his bond belongs, and a sale in foreclosure is likely to leave a sum sufficient to cover his mortgage, it would evidently be injudicious to resist "scaling down." Foreclosure would also result in loss, and only kill the goose that lays the golden eggs, in other words redeem the bond which yields the interest. Let us suppose that a road has a first 5 per cent. mortgage of $10,000,000, a second 5 per cent. mortgage of $10,000,000, earns $900,000 net, and is worth $18,000,000. Then it will have only $400,000 for interest on the second mortgage bonds, that is only 4 per cent. instead of the 5 it should pay. If in that case a plan of reorganisation is promulgated which proposes to scale down the interest to 4 per cent. it might be advisable for a second mortgage bondholder to give his assent; for, if the road was foreclosed, it would only fetch $18,000,000, a sum which would leave but $8,000,000 for the second mortgage, or $800 per $1,000 bond. Thus no gain would be possible, whereas serious loss might be averted; it being usual, in similar cases, to make up for the reduction by giving income bonds or shares to the "victims." But, on the other hand, it would in the above case be foolish for a *first* mortgage bondholder to assent to a reduction, unless there were some exceptional reason. He is safe. The earnings of the road amply cover his interest, and a sale in foreclosure would cover his principal; and as a matter of business he will therefore demand his pound of flesh.

Companies occasionally propose a reduction of interest without a reorganisation or without foreclosure. Whether the bondholder should assent or not naturally depends upon circumstances. If any material saving may be effected by resistance, let him refuse assent by all means; but not if the reverse is the case. It has been observed, as indeed is quite evident, that resistance is rarely offered in cases where principal and interest are badly covered by the value of a property or its earnings; and on the other hand a modification of the terms of a mortgage deed is rarely proposed when the reverse is the case, for it stands to reason that the proposal would have no chance of being accepted. The main point is that a bondholder can always force the debtor company to fulfil its engagements, if it is able to. All legal decisions clearly demonstrate this.

There have been one or two very interesting cases illustrating this principle. Some time ago the Minneapolis, St. Paul, and Sault Ste. Marie R.R. proposed a reduction of the interest on all its bonds from 5 to 4 per cent. : in return for this reduction the bondholders were to receive the guarantee of the Canadian Pacific, which, of course, would have raised the status and value of the bonds. A majority, therefore, consented; but one bondholder, belonging to the minority, protested, and threatened to bring a test suit; and we may take it as conclusive evidence of the unassailable merits of his case that the company did not allow it to come before the Court, and prefers to pay the full 5 per cent. to the dissenting bondholders. This is one of those cases which the companies are not at all anxious to have settled once and for ever by a Supreme Court decision : they prefer uncertainty to certainty. It is, by the way, the same with pooling. No company has ever permitted a pooling case to go to a Supreme Court; they know too well what the decision would be. But to return to the subject, in 1892 the Chicago, St. Paul, and Kansas City R.R. propounded a most high-handed scheme of reorganisation, and tried to induce its security holders to accept instead of their bonds some inferior securities of the Chicago Great Western, into which the company had been merged, The junior securities were unduly favoured by the scheme, and in consequence some large holders and a few small ones resented what they regarded as most iniquitous treatment, and intend to insist upon their rights. When the time for it comes they will go to law, and there cannot be the shadow of a doubt that they will carry the day, especially since the American Courts always display a tendency to uphold American credit.

Thus American railway mortgage bonds are, in the abstract, exceptionally well secured; but since they exist in great variety it need hardly be said that they are not all equally safe. Their great number and variety, however, furnishes the investor with an almost infinite choice : indeed, with the extraordinary diversity of their minor rights or claims, superadded to the difference in their fundamental character, one may say that scarcely two of them are alike.

Let us enumerate the principal varieties, and specify their chief characteristics. Perhaps it is desirable to premise a few remarks pertaining to their text, and we will therefore say that a bond generally acknowledges that the railway company issuing it owes a certain sum (usually $1,000 but sometimes $500, or, with sterling bonds, £200 or £100) to bearer, payable on a fixed date and at a certain place; it stipulates the rate of interest and the intervals at which such interest is payable; it states the amount of bonds belonging to its class issued, and specifies the property which is pledged as security for the payment of principal and interest (usually inclusive of the earnings of such property), and further gives such stipulations as to redemption by drawings and sinking funds, conversion into other bonds or shares, etc., as may be necessary in its individual case. Bonds are usually signed by the president and treasurer of the railroad company, and by the trustees to whom most of them are made out, and who must defend the rights of bondholders should the company fail to meet any of the obligations undertaken in the mortgage deed. This document enters still more fully into particulars than the bonds.

Next to this general feature comes the first subdivision, namely, the kind of money in which principal and interest are payable. There are: (*a*) sterling bonds, (*b*) gold bonds and (*c*) currency bonds. *Sterling bonds*, of course, are issued specially for the European market, and their interest being always payable in London they enjoy a certain distinguished position, enhanced by the fact that sterling bonds are usually prime securities anyhow. The distinction between *gold* and *currency bonds* dates from the war period. If there should ever again be a gold premium in the United States, currency bonds would of course be inferior to gold bonds, and although such a premium seems very unlikely, many people display an unmistakable partiality for gold bonds. As a curious illustration of this I may mention that the Alleghany Valley Railroad made an issue consisting partly of 6 per cent. gold bonds, and partly of 7 per cent. currency bonds.

After this division there is that which relates to

the nature of the property pledged under the trust deed. There are—

1. Ordinary mortgage bonds.
2. Equipment bonds.
3. Land grant bonds.
4. Collateral trust bonds.

Ordinary *mortgage bonds*, of course, are by far the most common variety. They have a lien on the entire property and its earnings, subject only to the rights of previous issues or to receivers' certificates or prior lien bonds (see below), but of course they can in case of foreclosure never receive more than is due to them under the stipulations of the mortgage deed. It happens quite often that a company issues *second mortgage bonds* which have a lien upon the same property as the first, but of course rank after them; hence they have usually a lower "credit." Some companies have issued *third mortgage bonds*, which naturally rank after the second; but to pledge a property up to the hilt in this fashion betrays invariably a very unsatisfactory state of finances.

These are the principal varieties of ordinary mortgage bonds, but there are some with additional epithets, notably "division," "extension," and "consolidated" or "general" mortgage bonds. These designations, however, have less to do with the rank of the bond than with the nature of the property they pledge. Thus *division bonds* have a lien only on a division or section of a road, and not on the entire property. If a company requires funds for extension it often issues *extension bonds* which have prior rights upon the new extension to be built with their proceeds, and frequently an additional lien upon the remainder of the property; the latter ranks, of course, after the rights of existing issues, for it would be impossible to grant a junior security rights already given to existing bonds; and it need hardly be said that the right of precedence once possessed by a particular bond can never be abrogated in favour of another except with full consent of its holders.

It happens frequently that various descriptions of bonds are consolidated or unified, and thus *consolidated* or *general mortgage bonds* are created. The issue of a general or "blanket" mortgage is usually resorted to with the double aim of obtaining new capital for extensions, improvements, etc., and of unifying various older descriptions, such unification being as a rule coupled with a reduction of interest, when this is possible under the stipulations of mortgages previously executed, or consented to by their holders. It is superfluous to remark that such reduction is never voluntarily acceded to by bondholders unless a company gets into straits and is reorganised; and hence parts of general mortgage bond issues are usually kept in hand until existing descriptions fall due, when they replace them. When a general mortgage is issued the amount in excess of the portion destined to retire older descriptions is used for extensions or betterments; in the case of extensions the bonds have prior rights upon such new parts of the system as are completed with their proceeds; in the event of betterments it is obvious that their rights must yield precedence to older mortgages resting on the property.

Apart from ordinary bonds there are, as we have already seen, several other descriptions. Concerning these we can be brief. *Equipment bonds* are issued to acquire rolling stock, and are secured by a mortgage thereon; and as rolling stock deteriorates in use, they are as a rule redeemed out of sinking funds. Akin to them are *car trust certificates*, not met with in this country nor quoted in New York, by the aid of which rolling stock can be bought on the "easy payment" system. These certificates are due at frequent intervals—usually semi-annually—and are secured by a lien upon the rolling stock in payment for which they are given. These car trust certificates are not always above suspicion, which is the reason why people do not like to invest money in them.

Land grant bonds pledge lands granted by the Government (see Chap. I.) as a guarantee for regular payment of dividends. The land being gradually sold, and the proceeds applied to the retirement of land grant bonds, the majority of these securities have been redeemed by this time. Owing to the low prices realised for land (usually less than $2 per acre), and considering the rapid rise in the value of real estate, it would probably have been more advantageous to railways to have retained their lands for some time to come, but need of funds and the advisability of reducing fixed charges usually induced them to accelerate these sales.

Collateral trust bonds are issued against no other security than funds of other railways.* Many of the greater companies have invested enormous sums in stocks and bonds of subsidiary concerns, mostly for the purpose of acquiring control; the Pennsylvania Railroad Company, for instance, possesses upwards of $150,000,000 (nominal) of similar securities. When new funds are required they are often obtained upon the security of bonds or shares of subsidiary concerns which are given in trust to trustees. The value of collateral trust bonds, like that of all mortgages, depends upon the value of the security upon which the advance is made, but usually the securities given in trust yield more than is required for the service of the collateral bonds, in which case the surplus as a rule goes towards a sinking fund out of which the trust bonds are redeemed.

We now come to those descriptions of bonds which differ from the ordinary kinds. These are:—
1. Receivers' certificates.
2. Prior Lien bonds.
3. Debentures.
4. Income bonds.

Receivers' certificates are rarely met with. They are issued by receivers for the sole purpose of keeping the property together, in other words, in grave emergencies only; and, for this reason, the Courts

---

* Sometimes also against securities of the issuing company.

have granted them first rights upon the property, and placed them in front of all other bonds. Akin to them are *prior lien bonds*, which are almost as rare as the certificates just mentioned, and likewise a sign of distress, either past or present, but usually both. They are, as a rule, issued in order to pay debts enjoying priority over the funded debt, and taken up by the holders of existing bonds who, however, must in any case be asked to consent to their obtaining precedence. But, as these bonds are always issued to provide a company whose credit is exhausted with urgently needed funds, this assent is gladly given.

Debentures are not frequently met with in America. They are bonds without any special collateral security, ranking after specified mortgage bonds, and their value depends entirely upon the financial status of the company by which they are issued. Thus in 1889 the Wabash issued $30,000,000 debentures which have thus far yielded no returns and which are practically issued upon faith, it being doubtful whether the Wabash, if sold to-day, would yield $30,000,000 more than the amount of its other funded debt—$48,000,000. These debentures will be worth little until the Wabash has materially improved its standing. The New York Central, on the other hand, has issued 4 per cent. debentures, which are quite a high class security, because the company has an excellent property and an excellent credit. For corporations similarly situated, debentures are probably the best securities to issue, because they give the capitalisation an elasticity which bonds cannot impart. But, in order to issue a safe debenture, a company must be quite first-class, and have a first-class credit. At the time when most bonds were created, these conditions were not fulfilled; and, moreover, the hard and fast rights of bonds tended to considerably enhance their value in the eyes of the investor.

No doubt debentures will gradually become more common in America, in the same degree as existing bonds mature, and as the credit of the companies improves; but, for the present, their scarcity is quite natural. Only five companies have, thus far, issued securities of this description; the two already mentioned, the Alabama Great Southern group which belongs to an English Company, the Delaware and Hudson R.R. and Canal Company, and the Chicago Great Western, a corporation of a very low order, which succeeded the Chicago Great Western after the latter had broken faith with its creditors in 1892. The last case is a curious freak; the others are due to unusual causes, with the exception of the New York Central and Delaware and Hudson, whose positions warrant the issue of a security entirely based on faith.

Income bonds more resemble preferred shares than bonds, no interest being paid on them unless earned. At the same time they are akin to debentures inasmuch as they are not issued upon any special security. They differ from preferred stock because no voting rights attach to them, and from debentures inasmuch as interest upon them in the majority of cases is non-cumulative. Income bonds are as a rule only issued in case of reorganisation, where bonds are "scaled down." Their holders then obtain an amount of "incomes" which will cause them to receive the old interest should it be earned. The creation of incomes was a prominent feature of the Atchison reorganisation, but this company exchanged them for second mortgage bonds when it had improved its position.

Guarantees are of such frequent occurrence that it is necessary to mention them, however briefly. Many of the larger companies have guaranteed bonds of subsidiary concerns, either as to principal or else as to interest; guarantee of both is still more general. It happens very often that the minor company is leased in consideration of such guarantee,* which relieves it of cares, and raises the status of the guaranteed securities; occasionally, too, a guarantee is given as the price of a traffic agreement, etc. Yet with regard to these guaranteed bonds it is somewhat curious to observe how their credit as a rule remains considerably below that of the guarantor company; and for this reason they seem to deserve the special attention of investors.

There are but very few American railway obligations which are perpetual, and most of them mature at a fixed time from the date of issue. The 100

---

* See reference to the case of the Minneapolis, St. Paul and Sault Ste Marie Railroad on p. 113.

and 50-year bond is quite common now, but 30, 25, and 20-year bonds are more plentiful, because they were issued at a time when an early improvement of the company's credit might be looked for. Nearly all bonds which become due at a fixed period are redeemable at par, but a few may be paid off at any time at the company's option, usually at a price slightly above par, say 105 or 110. As will be seen from the remarks preceding the list of bonds in the appendix, the time and conditions of maturity are of considerable interest to the investor; if a bond is purchased at a price exceeding its redemption value it loses, and if it quotes below that value it of course gains. Then there are some bonds that can be exchanged into stock before they fall due, like several descriptions of the St. Paul; and in these cases such rights, if worth anything, should be exercised.

When bonds fall due they are either paid off and replaced by others, or extended. The latter happens if the interest of the maturing issue corresponds with the company's credit, the former if its credit has improved. For example, a company which can borrow at 4 per cent. is not likely to extend a 6 per cent. bond except in those cases where the bonds are held by persons in control of the property. In such an event they may be extended without reduction of interest, although the course is dishonest; it amounts to robbing the company and the stockholders. But this high-handed procedure is very rare. Occasionally a bond is extended on condition that its interest is reduced.

In numerous cases there are *sinking funds*, into which a given sum has to be paid at stated intervals, the fund to be applied to redemption of a particular issue within a specified time, either by purchase at a fixed price, or else by drawings. In the case of drawings, the bondholder, when he calculates the yield of his security, must take into account his chance of having to surrender his bond at a price either above or below what he paid for it.

It should be noted that there exists a pronounced tendency to reduce the bonded debt of American railroads whenever this is possible. Of course, the smaller its funded debt the easier the position of a company. But as yet only a few corporations are engaged in this commendable pursuit; the Chicago and Alton and several New England roads being the most notable among them.

We now come to shares. As is the case with English shares, American railroad share certificates stand registered in individual names; but the practice of transferring them differs in the two countries. In England the transfer is upon a separate deed, whereas in America the deed, or power of attorney, is printed on the back of the share certificate; and the transfer being signed in blank by the registered owner, American railway shares are virtually bearer scrip. This practice has produced a system which has led to a great deal of abuse and inconvenience. Shares, especially those which pay no dividend, pass about for years in the name of the registered owner, although the latter may have entirely parted with them. Even in the case of dividend-paying shares the same difficulty presents itself, so that a person holding stock in the name of another has to produce the certificate to the registered owner, who collects the dividends.

This drawback is considered one of the reasons why shares are rather unpopular in England, especially in the provinces, and at various times attempts have been made to remove the objections arising from this source. In Holland, where great amounts of securities are held, the difficulty was emphasized by the fact that the majority of stock-owners were unfamiliar with the language in which the text of shares was written, and in consequence about half-a-dozen so-called "bureaux of administration," where shares are deposited against certificates, were established by leading houses; and these bureaux, having direct relations with the various companies, receive dividends in bulk, and distribute them amongst the shareholders, after deduction of a small commission for collection. It is evident that this method possesses great advantages, not the least of which is that shareholders, without much trouble, can be represented collectively, which results in their united voting powers being usually applied with the greatest effect possible under the circumstances.

The obvious advantages of such a system, namely accessibility of dividends and concentration of voting powers, resulted in its introduction into England being attempted some nine years ago, when the English Association of American Bond and Shareholders* was founded. To all intents and purposes

* A circular issued by the Association thus sets forth the disadvantages of the existing system:—

1. That the collection of dividends from nominal holders is attended with much delay, expense, and often with loss of the dividend, when the nominal holder cannot be found. In case of the death of the registered nominal holder, the collection of dividends through Trustees is almost impossible, and in all cases involves legal expenses. There is also the danger, which has recently been experienced, of Shares being passed on forged transfers.

2. In case of the death of an owner of American Railway Shares who holds the Shares registered in his own name, the transfer to executors can only be obtained by the original Shares being sent out to New York for fresh registration, together with a certified copy of the will, involving much delay, the cost of insurance, and legal expenses.

The advantages arising from registration of Shares in the name of the Association are summarised thus:—

1. On Shares registered in the name of the Association the voting power is combined, and the dividends are received by cable-transfer on the day of collection in America, and are at once distributed by cheque of the Association in London.

2. To avoid the risk of sending Shares through the post to be marked for Dividend every time a Dividend is payable, the Association is prepared to issue against Shares registered in its name the Certificates of the Association, countersigned by the London and Westminster Bank, Limited, against the original Shares deposited by the Trustees and Directors, lodged with the Bank. To country holders this is a great convenience, as the Certificates bear numbered coupons, and all that is necessary for a holder to collect his Dividend is for him to forward the coupons to the Association or to his Banker or Broker. The payments are notified very thoroughly through the press, thus, for example:—

The English Association of American Bond and Share Holders, Limited, will pay on and after [the day mentioned], Coupon No. 12 on their Certificates representing Shares in the Pennsylvania Railroad Company at the rate of three per cent. (or at whatever rate is declared by the Company).

3. The Certificates of the Association are all to Bearer, and for $1,000 each (either ten Shares of $100 each or 20 Shares of $50 each). In case of death these Certificates may be distributed by Executors without going to the expense, delay, and trouble of obtaining fresh registration on the books of the Railway Company.

Holders have the option, if they desire to do so, of retaining the original Shares of the Railway Company registered in the name of the Association, with Transfers signed in blank, and obtaining their Dividends on presentation of their Shares at this office to be stamped for Dividends claimed.

this association is identical with the Dutch bureaux, but owing to some opposition by other interests its certificates have not as yet been accepted on the London Stock Exchange, although the directors feel confident that in due course they will succeed in obtaining a concession to which no serious objections can be made and which elsewhere has been found to work well.

There are only two kinds of shares, namely, *common* and *preferred* stock. The latter, as its name implies, enjoys an amount of preference, varying in most individual cases, but always consisting of prior rights to dividends. Thus, to give an example taken at random, St. Paul preferred stock is entitled to 7 per cent., non-cumulative, after fixed charges have been met, before anything can be distributed on the common shares. The company has the privilege of reserving, as working capital, a sum not exceeding $250,000 over the floating debt, and the accrued interest of the mortgage bonds. After payment of 7 per cent. on preferred stock, both classes share further profits *pro rata*. The circumstances under which a dividend must be paid are, as a rule, most clearly defined, and there is not room for much discretion on the part of directors, except with regard to "betterment" expenditure, fully discussed in Chapter V. Preferred stock is usually non-cumulative, and hence, if it does not receive its full dividend in any given year, it has no claim to payment of the deficit in succeeding years. There are, however, a few instances of preferred stock being cumulative.* Concerning common stock, it is not necessary to add much to what has already been said on the subject. These shares simply get "what is left," and that being so it happens sometimes that they "get left" themselves.

After these details there remains only one more item to be mentioned. The chapters on capitalisation and the application of revenue have already fully dealt with the modes of issuing bonds and stock, the power to create new issues, and the principles affecting the distribution of returns amongst bond and shareholders. We now have only to state the rank of the claim which the various descriptions have as to payment of interest and principal.

After the payment of operating expenses, taxes, and rentals (that is "fixed charges"), the revenue goes to the various classes of stock in the following order. We omit collateral trust and land grant bonds because these, as it were, provide for themselves automatically. The securities then rank as under:—

1. Receivers' certificates.
2. Prior Lien bonds.
3. First Mortgage bonds.
4. Second   ,,
5. Third   ,,
6. Debentures.
7. Income bonds.
8. Preferred shares.
9. Common shares.

---

* For example, in the case of the Susquehanna and Western.

INTEREST.—DEFAULT. 129

Regular payment of interest on all bonds except incomes is of course compulsory, and so is redemption at the stated period. But some grace is usually allowed, as a rule six months. Occasionally default is permissible under the mortgage deed during a certain period; thus, to mention an example, Erie second mortgage coupons may be left unpaid for three years before their claims can be enforced. If a company cannot meet its obligations the bondholders can foreclose the mortgage, and sell the property. But usually the company obtains relief in some form or other when it gets into trouble; the coupons are deferred or funded, or the company is reorganised. Default on shares, it need hardly be said, is an impossibility.

CHAPTER VII.

SUMMING UP THE CASE.

The preceding chapters contain evidence which ought to enable every one to form in the abstract a fairly accurate estimate of the value of American railroad securities as investments. Nevertheless, it seems advisable to bring together the principal facts bearing upon the question.

Let it be understood, however, that generalisations applying to every kind of security are an impossibility. The United States is a big country, and her 3,000,000 square miles of territory embrace regions of all kinds and descriptions. Between the arid wastes of Arizona and the crowded New England States there range a series of communities passing through the most varied stages of development; between the fertile but untilled plains of Dakota and the "garden region" of New Jersey and Pennsylvania one finds numerous degrees and gradations of cultivation and progress. Here the country is exclusively agricultural; yonder it subsists almost entirely on manufactures; now it contains backwood regions, then highly-advanced districts; and necessarily the people in the various sections possess widely different characteristics, corresponding with their surroundings. As it is with the country and its people, so it is with their institutions in general, and their railways in particular. There are great systems and small ones, there are good roads and bad lines, there are sound companies and unsound enterprises. The Lake Shore and the New York Central are roads as perfect almost as any in the world; but in some sections of the South, in the woods of Wisconsin, in Kansas or Dakota there are lines which, from a technical point of view, must unavoidably be classed amongst the worst in existence. A number of companies, like the Alton or the Illinois Central, are corporations of the highest standing; others, like the Reading or Chicago Great Western, belong to the very lowest type. With the railway securities it is the same. There are not many better stocks than New York Central debentures, but there are only a few with less intrinsic merits than New York, Pennsylvania and Ohio common shares; yet the Erie, though an inferior company, has issued bonds which enjoy a very high credit. Between these extremes, technical and financial, one finds an almost infinite number of subtle gradations, connecting best and worst in the same way as the countless delicate tints of the spectrum connect red and blue.

It stands to reason therefore that generalisations can never apply to individual cases. Yet how often one hears phrases such as these: "All American securities are bad," or "No American securities are trustworthy." Especially when something goes wrong these phrases fill the air. When the recent Reading *débacle* took place, I was told by the editor of a leading daily newspaper that "These American railroad people are a larcenous lot, all of them." When Reading shares and incomes tumbled down, the *Economist** wrote: "Recent events have proved the delusiveness of the hopes that American railroad conditions have materially improved;" but that last expression of opinion was one of the soberest I found amongst a selection of some fifty editorial articles on the Reading affair and the wickedness of American railroad finance which it "proved." Now, it would be foolish to assert that American railway conditions are in any way absolutely perfect; but, nevertheless, similar generalisations are extremely absurd. Even if there are "larcenous" persons amongst railway people it does not necessarily follow that all are a "larcenous lot." And granted that there are scandalous events, that does not show that all conditions are rotten. No doubt there are some very bad companies, and some very bad securities. But that does not entitle anybody to arrive at the conclusion that *all* are bad. The European who says that no American securities can be good because the Reading and Richmond Terminal are bankrupt, or because the management of the Missouri Pacific bears little resemblance to Cæsar's wife, reasons exactly as foolishly as the Yankee who says that British Consols are bad because Uruguay is fraudulent and Portugal bank-

---

* 18th March, 1893.

rupt. The analogy is perfect, and the one argument not a whit sillier than the other. Undoubtedly there are such things as bad American railway securities; but in so vast a field of investment that fact detracts nothing from the value of the good stocks. And in this case as well as in any other it is the business of the investor to distinguish the chaff from the corn.

Let us now briefly recapitulate the main points to which this little volume has made reference. The relations between the railroads and the people, as we have already seen, have undergone a great improvement. The era of grievances against the railroads of the Republic has passed, or at least these grievances are gradually disappearing; here and there the Legislatures of new regions, it is true, seem all but friendly, but in the past it was the same in older States, and the experience there entitles us to forecast that the hostility in young States will not do much harm, and will ultimately disappear. The threatening attitude which six or seven years ago caused most railroad men to look forward with a good deal of apprehension had even wholesome effects for all concerned; like a thunderstorm, it cleared the atmosphere. Conditions are by no means ideal yet; railways continue to give some reasons for complaint, and a number of laws might be repealed with advantage to both parties. But gradual concessions are made on either side, and whatever strain there is will be eventually relieved unless new excitement is caused.

Among railways themselves more friendly relations have taken the place of the old hostility, and love of war has been driven back by a sincere desire for peace and harmony. Keen competition and rate wars have reduced tariffs to a phenomenal degree; but consolidations, amalgamations and traffic associations have stayed the decline, and wonderful technical improvements have reduced expenses to such an extent that profitable working is still possible. There is no reason to anticipate a further serious fall in rates,* and unless the expenditure of reasonable sums on betterments be prevented, such considerable reductions in operating expenses must be made that earnings per unit of freight will increase, while in addition traffic per mile will continue to grow. This growth would be prevented by a resumption of reckless construction, but a "craze" is unlikely to recur. Profits, however, must always depend upon efficiency, and anything preventing technical perfection is therefore to the detriment of railroad properties. Constant and judicious betterments are essential; every railway that has achieved greatness succeeded chiefly because it constantly aimed at perfection. To keep abreast of the times is the only sound policy of an American railroad. Fortunately all companies act upon this principle if they can, and the American bids fair to become the standard railway of the world. It advances by leaps and bounds, while those of other

---

* Whilst correcting the proofs I notice that this view is shared by Mr. Joseph Price; see a lengthy interview with that gentleman in the *Financial News* of June 15th, 1893.

countries stand still in the firm but erroneous belief that they have reached the goal of perfection.

Perhaps in no respect have changes been so healthy as in the relations between the corporations and their shareholders. To be sure, irregularities still occur, though mostly in the case of minor independent companies which happily are little known in this country. But the old era of "railroad rascals" has gone, and men of integrity are filling the vacated places. The American railway has ceased to be chiefly a gambling implement for Wall Street, and properties are no longer wrecked for speculative purposes. Swindles, "deals" and "steals" are not frequent now-a-days, and reorganisations and receivers are much less common than in the past. Though cliques still rule in many cases, they no longer tyrannize, and the great improvement in business morality, an improvement attending that rapid solidification of business upon which the country may well congratulate itself, has purified business methods and propagated honest and healthy financing. Arbitrary and fraudulent issues of shares and bonds for example, though formerly of every-day occurrence, are no longer heard of now, and construction companies, dishonest leases, etc., have become very rare. Greater attention is ever being paid to the stockholders and their wishes, and unjust treatment is no longer common; some time ago the shareholders of the Illinois Central were even consulted upon some important matters pertaining to extension,*

* In 1892, before the stock of the Louisville, New Orleans and Texas Railway was purchased.

a step, I believe, never taken before. The investor is kept better informed than he used to be, and reliable official statements are issued frequently and regularly. Matters of policy are discussed at length in annual reports, and the latter are so exhaustive that they might serve as an example for those issued by English railways. Publicity is aimed at and promoted; it is no longer feared, and this, perhaps, is the strongest proof of the rapid strides railways are making towards ideal conditions. A few dark spots, do, of course, remain, one of these being the arbitrary power most directors have over expenditure and dividends, and the abuses this may lead to; but such abuses are but rarely committed, and when a few "gentlemen" of the old school who conduct railway business upon old "principles" are gone, they will no longer be heard of.

In short, progress is such as could not have been anticipated ten years ago. There are companies now which in every respect are as good as the best English concerns, emphatically just as good in every respect; and those who do not yet belong to that class are trying with might and main to reach it at the earliest possible moment. But, in spite of this far-reaching amelioration, there is still room for improvement. There are a few large, and quite a number of small, companies which as yet have no management deserving the unqualified confidence of investors. To this class belong most Gould roads, the Southern Pacific group, the Northern Pacific, the Chicago Great Western, and it seems also the Reading, though some people have long hesitated to accept the now common version of Mr. McLeod's *régime*. In most instances this distrust is chiefly based on past history, and in many cases it is exaggerated. But it cannot be denied that there is some reason for it, and all we can do is to regret that the process of purification has not yet been entirely completed, and to hope that those roads which could do with a " Pentecostal visit of the Holy Spirit of Honesty," will soon follow the good example set by the better class. There seems little chance that we shall be disappointed in this hope. Most shady concerns are as yet comparatively young; and experience shows that in the American railway world as well as in some other spheres of life respectability usually comes with maturity.

Another point which tells against American railways is the vast amount of discretion directors are endowed with; perhaps I should say the abuses this discretion may lead to, and the feeling of insecurity with which it is apt to inspire investors; for especially in America the management of companies must always be more or less autocratic, since the investor, as a non-expert, must necessarily rely upon his directors, the experts. But experience shows that this inevitable faith is sometimes misplaced; and this possible abuse of immense discretionary powers constitutes the dark side of American railroad investments. It is *the* great point calling for the close attention of the investor. And, though the status of his company is naturally of considerable moment, the safety of his investment depends upon the discretion of the directorate, and the greater or smaller probability of an abuse of their discretion is really the principal question before him. He can only be really safe as long as he remains altogether independent of this discretion; but, where he cannot or will not be on the safest side, he should see to it that he does not entrust himself to men likely to put their authority to wrong uses.

But nature, as a rule, provides all defects with their counterpoise. The polar bear is white that his colour may not contrast with the Arctic snowfields: the scorpion carries in its body the antidote for the poison of its own sting. And the principal defect of American railway management has likewise engendered its own remedy. It has led to the prevalence of securities which are shielded against its possible consequences as far as human ingenuity can shield them. The American railroad bond of the better class is perfectly protected against possible abuses, and only a grave and general disaster can depreciate its value. In consequence it has gradually gained high esteem and become extremely popular as a medium for investment, not only in America and England, but also in Holland, Germany, France and Belgium. And it fully deserves this position.[*]

In the main, money should be invested in funds

---

[*] The following six pages are quoted from my larger work on American Railways.

possessing four cardinal qualities. Safety is the first and foremost of these; in the second place they should pay a good and regular return upon capital; thirdly, they must be marketable at all times without difficulty and without severe loss; and, fourthly, they should possess stability, or better still, their value should be progressive. If, therefore, we wish to estimate the value of American railway securities as investments, we must see whether they answer these four requirements.

First, as to safety. It is not difficult to find data showing the degree in which they possess this quality—of course in general; it need scarcely be said that the present remarks cannot be anything but general. On January 1st, 1891, the latest date for which statistics are available, there were in the United States 170,601 miles of railway. Their total funded debt, according to *Poor's Manual*, amounted to $5,235,000,000, or $30,687 per mile, a price equal to what is generally assumed to be the average real cost of American railroads. But, as we have previously shown, the total bonded debt is really considerably below Mr. H. V. Poor's aggregate on account of the duplication of stock (p. 137); and, in reality, railway property is probably not mortgaged beyond the extent of $24,000 per mile. Hence the property upon which bonds are issued leaves an ample margin for depreciation, however little reason to anticipate the latter there may be. Its earnings, too, amply cover the interest; and from this it follows that bonds, broadly speaking, are quite secure both as to principal and interest.

A desirable investment should also pay a good and regular return upon its cost. On the average American railways, according to Mr. Poor, pay 4·25 per cent. on their bonded debt, but, on account of Mr. Poor's divisor being too large, as we have pointed out before, their real returns are considerably in excess of this ratio. If we assume railway property to be mortgaged to the extent of three-fourths of its apparent bonded debt—a low estimate—bonds give an actual average of 5·67 per cent. per annum (an estimate tallying with a previous calculation*) according to which some high-class descriptions quoted in Europe pay somewhere about $5\frac{1}{2}$ per cent. upon their face value. This average must be called exceptionally good, and its regularity leaves nothing to be desired, there being very few mortgage bonds indeed the returns from which are irregular or uncertain.

The third requirement is also fulfilled. Bonds are marketable at all times, and, as is shown by the following tables, the quotation of the better class of bonds is subject to changes exceedingly insignificant, both in themselves and in comparison with the fluctuations to which the price of most shares is liable.

---

* See p. 138 *American Railroads as Investments*.

Table showing the average price of various American Railroad Bonds dealt in in London for each of the five years ending with 1891.—New York quotations.

	1891	1890	1889	1888	1887
Baltim. & Ohio 5 p. ct. Gold Mortgage	105¼	107½	109	108	106¼
St. Paul First 7 p. ct. Gold ,,	124¼	124¼	125½	125½	128½
Ch. & Northw. 5 p. ct. Sink. Fund ,,	108¼	115½	109½	108¼	108
Illinois Centr. 4 p. ct. Gold ,,	102½	101½	107½	106½	107¼
New York Central 7 p. ct. 1st ,,	126	129	135	134⅜	131⅞
Erie 7 p. ct. 1st Consol ,,	135	134¼	135¼	135¼	133
Pennsylvania 4½ p. ct. ,,	105	107¼	107½	107¼	105¼

According to this compilation, the annual fluctuations of these bonds are as follows:—

Table showing the difference between the highest and lowest quotations of various American Railroad Bonds dealt in in London for each of the five years ending with 1891.—New York quotations.

	1891	1890	1889	1888	1887
Baltim. & Ohio 5 p. ct. Gold Mortgage	6¼	5¼	4	6	12½
St. Paul 1st 7 p. ct. Gold ,,	8¼	4½	3	5	7
Ch. & Northw. 5 p. ct. Sink. Fund ,,	6½	3	5	5	5
Illinois Central 4 p. ct. ,,	5	5	5	3½	3¼
New York Central 7 p. ct. 1st ,,	5	6	5	4¼	5¼
Erie 7 p. ct. 1st Consol ,,	5	8	5½	4½	8
Pennsylvania 4½ p. ct. ,,	6	5	4	4	4

These tables show that the price of first-class bonds is remarkably stationary. The annual fluctuations of the bonds quoted above are, with one exception, smaller than those of London and North-Western stock, which yields much less nett. As regards the value of bonds usually classed lower, it is true that fluctuations are wider. But this is quite natural, and in most instances there has been a decided rise in prices during the past five or six years, and their "credit" has undergone a considerable change for the better. Nor can this be wondered at. In the first place outside influences have of late been decidedly in favour of securities yielding good returns, but apart from favourable extraneous influences American bonds have benefitted by an amelioration of their intrinsic worth. The value of bonds, especially of those not as yet ranking among the highest class of investments, has risen rapidly, and this rise is fully warranted by circumstances. There has been an upward movement in the average value of railroads; improvements have been effected, and powers, comparatively speaking, enhanced. Honest management has superseded a *régime* of rascals; and although the country has by no means reached maturity, while business is as yet not entirely weaned from some doubtful practices, there is a vast improvement in commercial morality, and a process of solidification has set in which it is very gratifying to observe. Beyond these favourable changes there is the auspicious fact that the rapid growth of the country must react upon railway business and render the properties more productive and therefore more valuable as time goes on, unless reckless construction is resumed, an event which may be regarded as very unlikely. For these reasons it not surprising that we entertain a higher opinion of American railroad bonds—and especially of the lower grades—now than we ever did before. Throughout the recent crisis Europe has been buying American bonds, though it has sold large blocks of shares.

To what extent their value is appreciated has been shown by previous tables, and is further manifested by the fact that most bonds of responsible companies are quoted above par, whereas they were sold below. Eleven leading companies, already referred to, pay an average interest of 5·54 per cent. on their entire bonded debt; but the public buys their securities at a price at which they yield less than 5 per cent. Yet this figure is so high in comparison with the net returns of other investments that the question arises whether the credit of American railroad securities does not occupy a level lower than that which might be properly conceded to it in England.

The *Investors' Review* used to contain a valuable appendix showing the net returns upon thousands of investments, and by kind permission of its Editor I compiled from it a table showing the average net returns paid by various classes of railway securities quoted in London, which enables us to see what credit American railroad securities enjoy in comparison with other railway stocks quoted in our market. The table was made out in May, 1892, and was a laborious piece of work; and as the change in prices which has since taken place does not affect its value in any material way, I simply reproduce it without bringing it "up to date."

Statement showing the comparative returns upon investment paid by 532 descriptions of railway stock according to London quotations of April 15th, 1892. (Compiled from the Investors' Review.)

RAILWAYS.	Less than 3 p.cent	3 to 3¼ p.cent	3¼ to 4 p.cent	4 to 4½ p.cent	4½ to 5 p.cent	5 to 5¼ p.cent	5¼ to 6 p.cent	6 to 6¼ p.cent	Above 6¼ p.cent
British (all descript.)	6	185	45	11	3	1	—	—	—
Indian ,,	—	17	7	3	1	—	—	—	—
British Possessions ,,	—	—	5	20	5	—	4	4	—
American (bonds)	—	1	15	39	42	25	10	8	1
South American......	—	—	1	4	13	10	—	6	—
French and Belgium..	—	3	3	—	—	—	—	—	—
Balkan States........	—	—	—	—	4	2	2	—	—
Other Continental....	—	—	—	3	6	6	3	—	8
Total	6	206	76	80	61	44	19	18	9

This table indicates that the vast majority of English railway stocks *of all descriptions* are deemed worthy of a "credit" ranging between 3 and 3½ per cent., but approaching the lower of the two figures, and that it is the same with Indian railways. French and Belgian railways are evidently thought just as well of; colonial lines mostly pay between 4½ and 5 per cent.; the bulk of American bonds are estimated to deserve a credit of between 4 and 5½ per cent., the same figure as sundry descriptions of South American and Balkan States companies, usually enjoying Government guarantee. It is especially this last fact which seems to denote that their credit is under-estimated; and if we examine facts more closely this supposition appears to be confirmed.

According to the periodical just quoted (May number, p. 342) there are but twelve descriptions of American bonds, issued by six companies, returning between $3\frac{3}{4}$ and 4 per cent. on the net investment. They are the very best bonds issued or guaranteed by the very best companies, namely the Pennsylvania, New York Central, Illinois Central, Boston and Maine, and Chicago, Milwaukee and St. Paul Railroad Companies. They have prior rights upon properties all of which, except one, are fully equal to the average English railroad, and never since they came into existence have any of these bonds failed to pay their interest regularly, nor is there any prospect of their ever doing so. To all intents and purposes, therefore, they are as safe an investment as any English railway share, and perhaps I might say safer, because there are people to whose judgment great value may be attached who maintain that sooner or later the imperative demand for lower rates will prevent many of our extravagant English companies from paying a dividend at all. Yet it is evidently thought that the pick of American bonds should pay £0 15s. 0d. per cent. more than the average English railway stock. With other securities, paying a higher net return, the contrast is still more marked. For example, Baltimore and Ohio consolidated 5 per cent. mortgage bonds quote 115, so that they return £4 6s. 10d. net; and Louisville and Nashville 6 per cent. general mortgage bonds 120, in consequence whereof their credit is no better than £4 16s. 11d. per cent. There is no guaranteed stock of any British railway which yields so much net, and there are but three kinds of ordinary *shares* (Belfast and County Down, Highland, and Midland Great Western of Ireland) whose credit is estimated at so low a figure as that of the Louisville and Nashville *bonds* just referred to. This is so altogether out of accord with intrinsic merits that it is absolutely unnecessary to call attention to it. Among others there are at present more than forty descriptions of American bonds quoted in London all of which are issued by companies who have never defaulted, and are certain to pay their interest regularly; and yet these bonds give as large a return upon net investment as, and therefore in our estimate of their credit are classed with, a miscellaneous lot of second-class foreign railway stocks—Turkish, Argentine, Central American, and so forth—which can no more rank with St. Paul, Northwestern, or Burlington bonds, than can the dilapidated Ottoman Empire or a turbulent Central American Republic with the United States. Similar anomalies are so striking that to enlarge upon them is absolutely superfluous. The most cursory comparison with other classes of securities must clearly show that the credit of American bonds is as a rule from $\frac{1}{2}$ to 1 per cent. below the point to which they are fairly entitled.

Presumably I am expected to state why they have no more. Two reasons suggest themselves. In the first place we underestimate their safety as a result of the lack of confidence which remains from the days when we were duped, and to the exaggerated nature of which I have referred before. Secondly, the fact that the price of bonds is regulated in New York rather than in London influences their quotation. American securities are mostly held in America, and there people are not satisfied with the low interest upon a safe investment that is acceptable to us. Houses and mortgages upon real estate often yield 8 per cent., and mostly between 5 and 7; and the fact that in America money has a greater value than here cannot but have an influence upon the price of investments the worth of which, broadly speaking, is fixed by Americans.

This last circumstance places the European investor at a great advantage which to some extent seems to be realised; at least our holdings increase steadily. But it does not appear that the advantages offered by this class of securities are understood to their full extent; if they were we should probably seize them with even greater avidity than we do now. There are at the time of writing some forty descriptions of American bonds paying between 4 and $4\frac{1}{2}$ per cent. upon net outlay, which, as a safe investment, find no parallel among any other stocks quoted at corresponding prices. To have the same return upon his capital an investor would, if he were to limit himself to railway stock, have no other choice than a very limited assortment of South American or Turkish railway "securities," or shares of Colonial or Home railways of very doubtful rank. Upon an investment in British, Indian, or Continental railway stock, possessing the same essential qualities, the return would be at least $\frac{1}{2}$, and probably 1 per cent. less.

Bonds, therefore, are all right, and, barring a few inferior descriptions, eminently fitted for investment. I believe even that the better kinds are more eligible than most other classes of securities, because they offer a much higher yield, whilst the risk is virtually the same. And it seems that others share this view; at least I know of a Scottish fortune of £6,000,000 made in America which is entirely invested in American railway bonds.

But there are other securities than bonds; and preferred and common shares necessarily do not occupy the same station as obligations, simply because they lack their safeguards, and rank below them. Yet they are looked upon with favour by people fond of "speculative investments," and of speculation pure and simple. When carried beyond legitimate limits, as it usually is, this tendency degenerates into a social evil, and for that reason " speculation," in the sense usually attached to it, deserves the most unqualified disapproval. But in its better and purer sense, in other words as long as it is guided by caution and keeps itself within legitimate bounds, the tendency towards speculation is, on the whole, not a bad tendency. Of all things, it is the pluck to take risks for the sake of gain which has made this country great and prosperous, as it has so many others. And there is no reason why people should altogether avoid "speculative investments,' provided they exercise due caution. American shares usually fluctuate widely within a very short space of time; the highest and lowest points touched each year have a broad chasm of many "fractions" to separate them, and I do not see why monied people should not buy at the low level, take their stock off the market, and "hold for a rise." Provided a man does not go beyond his means and exercises caution, he will, as a rule, find it possible to realise a handsome profit, even if a considerable decline takes place after his purchase.

Table showing the difference between the highest and lowest prices of nine descriptions of American railroad shares during each of the six years ending with 1891.—*London quotations.*

	1892	1891	1890	1889	1888	1887	1886
Central Pacific	8½	8¼	11⅛	3½	10¼	17⅛	13½
St. Paul..................	8⅞	32¼	36½	14⅜	18⅞	25¼	17⅜
Denver & Rio Grande....	3½	7¼	8¾	3⅜	8½	12⅜	15¼
Lake Shore	14	17¼	11⅛	9½	18½	9¼	26¼
New York Central	11¾	22⅝	15¾	5⅜	8⅜	13⅜	19¼
Norfolk & Western	9	12¼	23	15⅛	16⅞	20⅞	34½
Pennsylvania (2 X $50) ..	10¼	16¾	20½	10	10	14¼	17¾
Reading (2 X $50)	25	18½	21¼	13¼	11⅞	39½	30¼
Wabash Preferred	5½	17¼	11⅞	10⅛	8¾	14⅞	26⅜

The only trouble with speculation is that it degenerates so easily into gambling. One never knows where to draw the line with it, and it is impossible to say where it ceases to be legitimate, and becomes something worse. The term has, alas, become synonymous with gambling, and taken in that sense speculation in American railways is the same as speculation in other stocks; a dangerous practice which no responsible writer will encourage.

For that reason I enter but reluctantly into a question which, as can be inferred from the title of this volume, does not really form part of my present subject. Next to mining shares the lower types of American securities are, in consequence of their violent, sudden, frequent, and often unaccountable fluctuations, grist for the gambler's mill. Nevertheless I feel that I may not wholly ignore this side of the question, simply because it is so closely related to investment. And looking upon it from the practical point of view, one must recognise that even with regard to speculation matters are better than they used to be. The almost general abstinence of managers from speculating on a vast scale, and the gradual abolition of the once universal practice of making money in Wall Street by manipulating properties regardless of the interests of their owners, can have exercised none but a wholesome influence even upon speculation. But these have not been the only changes which have influenced "the Street," and there have been other developments which presumably are not regarded with favour by the speculator. The activity of hundreds of petty managers has been replaced by the vast operations of comparatively few leaders, and instead of being ruled by thousands of small powers, Wall Street has come under the influence of a few great operators, each of whom is blindly followed and supported by a host of nonentities. This change in the tactics of Wall Street could not fail to have far-reaching effects upon

speculation in American railway shares. Upon the
bonâ fide investor the new customs of "the Street"
have had no adverse influence; he benefitted by
the abolition of old rascalities perpetrated at the cost
of his property, and the new mode of speculation had
very little interest for him. Not so with the professional
"operator." Instead of hundreds of minor powers
whose actions frequently afforded a clue to their
intentions, and who often took him into their confidence, a few men gained control of the market and
ruled its destinies in an unaccountable way. In bygone
years the tactics of Wall Street had at least general
and comprehensible characteristics; at present they
have none, and the speculator is at the mercy
of men whose intentions are enshrouded in mystery,
whose power is almost unlimited, and whose tactics
are rarely understood. So great is the power of a
few of these leaders that they have pretty well
absolute control over the prices of speculative stocks,
and can move them up and down almost at will.
Formerly actual value regulated quotations, and to
alter the quotation of shares it was frequently
necessary to change the worth of a property first.
But to-day Wall Street takes little account of the
actual value of properties. There can be "booms"
when railway business is slack, and in prosperous
times dulness may prevail; and this well-known fact
clearly proves that actual value has little to do with
the price of most shares, and that the real condition
of railway business is of little use as a guide to
speculators.

The speculator cannot have recognised this last fact with great satisfaction. As it was, the complex nature of railway business in the United States rendered discernment of its true character and immediate prospects—and such discernment, after all, is the basis of speculation—a matter of great difficulty. To judge of the merits of vast systems, traversing regions with the most varied characteristics and running from one extreme end of the country to another, to estimate their earnings under varying conditions of trade, has at all times been difficult; but even those who can do this gain no longer any advantage from their knowledge. The great operators who make or mar "markets" can fall back upon resources that will set at bay any natural influence and baffle any amount of knowledge. The intentions of these great wire-pullers cannot be divined by the ordinary speculator; even the American, with his innate tendency towards speculation, has clearly recognised this. Were it not for the fact that in the human breast—especially with Americans—the hope of gain proves stronger than the fear of loss, Wall Street would have no outside support whatever. As matters stand it retains some, but it cannot be questioned that speculation in railway stock has lost much of its popularity as far as America is concerned. And those who abandon the sound principle "not to speculate in what you don't understand" speculate in a modest way, and in very much the same manner as that in which they buy tickets in the Louisiana

Lottery: they do it secretly, and although they hope to get a prize they know of the great number of noughts. Speculation in the United States, in spite of its remarkable development and its hold upon the people, no longer favours Wall Street to such a marked extent as before, and chiefly seeks other fields than railways, notably real estate.

CHAPTER VIII.

HOW TO INVEST.

So, in theory, money may be invested in American railroad securities with safety and advantage, provided people exercise due caution. But what is due caution? Under what circumstances may a person be said to be sufficiently careful? You will say when he sees to it that his money is placed in sound securities of sound companies; but that brings us little or nothing beyond the commonplace. People know all these golden rules. The only trouble is to make them live up to them, and to teach them to discriminate between a good company and a bad one, and between sound securities and rubbish. When you have taught them that, you have done the best you can, but the probability remains that many a man who knows all about these good principles will forget them when big "yields" and low prices tempt him, and come to grief in consequence.

We doubt whether it is possible to give information that will enable a person to form his own opinion concerning the soundness of a railway company, or the safety of a given description of stock. But at any rate we shall try to give a few useful hints.

The standing of a company is measured chiefly by its balance sheet and its income account; and if these are given in the "comparative" form usually adopted in America, that is with the rows of figures relating to a series of years side by side, it is possible to make some most useful deductions and comparisons which will suffice for most ordinary purposes. Of course, a look at these figures does not constitute the entire study; that requires some technical knowledge and a good deal of practice.

The reports of nearly all large American railway companies are so exhaustive, and compare, as reports, so favourably with those issued by English companies, that one may scarcely mention the two side by side. They state the cost of fuel, the number of ton-miles of freight, the average rates, the work done by every train and every engine, the cost of moving freight and the profit per unit, the composition of the traffic, the nature of the track, in short, the fullest details concerning everything. In consequence they enable the expert to draw a vivid and faithful picture of the position and progress of the company in a way that would be absolutely impossible in the case of English railway companies who, it seems, have no faith in such a high degree of publicity. But these details are of little use to the average investor, and the balance sheet and income account fully answer his purpose, the former chiefly by showing the status of the company, the latter by illustrating the soundness of its several securities.

The usual form of the (condensed) comparative balance sheet is something like the one given on this page. It is compiled from the two last annual reports of the Philadelphia and Reading Company, and therefore is of some additional interest just now; looking at the two rows of figures representing

Condensed Comparative Balance Sheet of the Philadelphia

Dr.	1891	1892
Capital Accounts—		
Railroads, wharves, and terminals	$80,261,112	$80,673,510
Locomotives, engines, and cars	18,666,710	21,702,772
Real estate owned	4,920,698	7,088,103
Steam colliers and floating equip. owned	1,291,243	1,707,429
Investment of R.R. Co. in Coal and Iron Co., represented by stock, bonds, and accounts	69,453,181	70,633,446
Railroad and canal leases under reorganisation plan (Schuylkill Nav. Co., Susqueh. Canal Co., Colebrookdale R.R. Co., Pickering Val. R.R. Co.)	9,341,161	9,341,691
Bonds and stocks owned by Company	22,669,000	25,984,825
Bonds and stocks of leased lines deposited with Penn. Co. for Ins., etc., trustee	2,466,777	2,466,777
	$209,069,882	$219,598,553
Current Accounts—		
Cash on hand	961,873	1,063,052
Bills receivable	104,613	59,506
Freight and toll bills	1,004,701	2,162,949
Materials on hand	801,306	3,352,071
Leased and controlled companies	2,615,874	4,691,025
Connecting railroad companies	370,499	1,136,516
Current accounts	298,878	566,324
Total	$215,227,625	$232,629,998

"current accounts" the remarkable augmentation in the floating assets and liabilities which was the cause of the recent troubles will at once strike even a non-expert.

and Reading Company, November 30th, 1891 and 1892.

Cr.

	1891	1892
Capital Accounts—		
Bonded debt prior to incomes	$77,293,936	$79,775,466
First, second, and third income bonds	58,099,797	58,376,706
Bonds and mortgages on real estate	2,052,090	3,490,496
P. & R. Terminal RR. Loan, 1891-1941	.. ..	7,258,000
Car trust certificates	3,906,000	3,174,000
Equipment notes	3,485,307	6,059,623
Sinking fund loan, 1892-1902	.. ..	1,985,000
Debenture loans	4,405,100	8,100
Common stock	40,105,362	39,830,362
Deferred income bonds (nominal par $25,568,090)	7,670,427	7,670,427
Miscellaneous	19,860	19,760
	$197,037,880	$207,647,939
Current Accounts—		
Notes payable and loans	1,150,000	3,412,567
Leased roads and canals, account rent	239,077	145,835
Interest due and uncollected	750,007	771,046
Interest and rentals accrued, not payable	1,336,725	2,861,153
Connecting railroad companies	235,548	265,462
Current accounts	208,735	370,940
Audited vouchers and pay rolls	1,975,836	3,930,624
Taxes	361,001	453,715
Surplus for year ending Nov. 30th	1,864,962	3,157,147
Balance carried to and held for account of future adjustment and suspense accounts	10,067,855	9,613,574
Total	$215,227,625	$232,629,998

A balance sheet contains two kinds of entries; the principal belong to "capital accounts," that is, what might be called the fixed status of a company; the others relate to "current accounts," which embrace far more elastic items. The former are of very trifling value and indicate little. On the "Dr." side, that is among the assets, we find the various properties, roads, rolling stock and investments; and on the credit side there are the liabilities against these properties, the bonds and the stock. The liabilities are as a rule a few millions below the assets, but these data say very little nevertheless. In the case of the Reading, for example, how are we to know whether "Railroads, wharves and terminals" are worth $80,673,510, and the "Investment of Railroad in the Coal and Iron Company" $70,633,446? They are not; but how should an investor see it from the accounts? The statement of current assets and liabilities on the other hand, is an extremely valuable indicator. It shows whether the company can transact its current business with facility or not. The Reading, for instance, was at the end of the two fiscal years 1891 and 1892 in the following relative position; we obtain these figures by addition of the various items under "current accounts" on the two sides of the balance sheet.

	1891.	1892.
Current liabilities	$6,296,929	$12,211,339
Current assets	6,157,743	13,031,445

Excess of current *liabilities* over assets in 1891 $139,186
Excess of current *assets* over liabilities in 1892 .... $820,106

Here we have what looks like a favourable change from 1891 to 1892, for in the latter year the current assets exceed the current liabilities by a small sum. An expert would at once proceed to further investigate the matter,* but our purpose does not render it necessary to enter into a similar analysis.

Since November, however, considerable changes have occurred. The company took to borrowing in various ways, and in February, 1893, its floating liabilities reached a total of $18,472,838, against which there were only assets to the extent of $15,779,784. Moreover, whereas the bulk of the liabilities were due almost at once, the assets were largely "locked up," consisting as they did of coal, book debts, etc., Thus the company became embarrassed; it could not meet its current obligations out of its current income, and went into the hands of a Receiver. It has a floating debt of huge dimensions, and it cannot get into smooth water until this floating debt is paid off or changed into funded liabilities, and until its current assets are liquid, and large enough to cope with the current liabilities against them.

Now let us look at the balance sheet of a sound company, say the New York Central; we shall then at once see the difference. The assets and liabilities of that company on June 30th, 1891 and 1892, were as shown below :—

---
* He would see what part of the liabilities are pressing, and what proportion of the assets is available at once; and if the urgent needs are not offset by *liquid* assets, such as cash in hand, amounts receivable, etc., he would immediately perceive that the financial condition of the company is not an easy one.

GENERAL BALANCE, N.Y.C. & H.R.RR., JUNE 30.

Assets.

	1891. $	1892. $
Road and equipment	151,002,288	153,585,294
Special equipment	5,706,464	5,406,464
Stocks and bonds	10,034,635	9,394,322
Ownership in other lines, real estate, etc.	4,169,701	4,568,929
*Due by agents, etc.	5,280,791	5,028,220
*Supplies on hand	3,072,813	3,337,891
*Cash	2,896,277	1,368,609
*Harlem construction account	1,049,984	1,263,541
*West Shore cons'tion acct.	....	643,433
*Miscellaneous	140,230	165,084
Total assets	183,353,178	184,761,787

Liabilities.

	1890-91. $	1891-92. $
Capital stock	89,428,300	89,428,300
Funded debt	65,377,333	68,077,333
Real estate mortgages	357,000	342,000
Securities acquired from leased lines	3,359,700	2,827,200
Past-due bonds	4,790	4,790
*Interest and rentals accr'd	3,890,039	3,660,211
*Unclaimed interest	14,324	11,089
*Dividends	894,283	1,117,854
*Unclaimed dividends	30,075	30,650
*Wages, supplies, etc.	3,822,833	3,544,994
*Due other roads, etc.	2,939,003	1,546,819
*West Shore constr'n account	9,472	....
*Rome, W. & O. cons. account	......	811,199
*Profit and loss	13,226,026	13,359,348
Total liabilities	183,353,178	184,761,787

The items marked with asterisks constitute current assets and liabilities. They foot up (1892) as follows: Current assets, $11,806,778; current liabilities, $10,723,876; and the current assets therefore exceed the current liabilities by more than $1,000,000. But that is not all. The assets are *liquid*, which, as we have seen, was not the case with the Reading. There is $1,368,000 in cash, and $5,028,220 " due by agents," the equivalent of cash, while the "notes payable and loans," so conspicuous in the case of the Reading, like other urgent liabilities, are absent. Thus a comparison of current liabilities and assets will at once show the difference in the position of the two companies. The New York Central, having no current liabilities except those inseparable from its business, and a large amount of available cash to meet these, was in a sound state: the Reading, with bills and loans due, and little cash or equivalent of cash, was in a precarious position. A company can, as a rule, only be of sound standing when its floating assets exceed its floating liabilities, and are liquid. It is not of sound standing when the reverse is the case, or when its current liabilities are urgent. Loans, etc., amongst these liabilities are nearly always a signal of distress, or, at any rate, denote financial weakness. A company should never have occasion to include amongst its current liabilities anything but items inseparable from its current traffic business. Notes, bills, and the like, are always suspicious. They constitute the floating debt, which must either be redeemed out of earnings or else funded, and always betrays some defect or other in the management.

Hence these "floating" items, and especially a comparison of them with the corresponding figures of previous years, give the investor some very valuable indications. But not so the other entries. Of course, bonds and stocks are issued, and constitute liabilities, and they are offset by property in some form or other.

But it stands to reason that the balance sheet will always show the value of this property to exceed, or at least to equal, the liabilities incurred in order to pay for it, in other words, its stocks and bonds. But how is the investor to know that this "book value" is real worth? Take the Reading balance sheet for example. It says the railroads, wharves, and terminals are worth (1892) $80,673,510, and computes the value of the investment in the Iron and Coal Company at $70,633,446. These figures are mere fables, as, in this instance, everybody happens to know. But how is the investor to say in every case to what extent analogous data can be relied upon in the case of other companies?

It is quite evident that it would be impossible to ascertain this value without either an examination of the property by an expert, or a thorough analysis of the company's report. But both means are not accessible to the average investor. He has either to rely upon the usually perfunctory and unreliable discussions of the newspapers, or else upon some golden rule or other. Of these rules there are several, but not one of them is good enough to deserve any faith. It has been said that the bonds and stock of a company, whose property is worth what it is capitalised at, will stand in a ratio of about 10 to 1 to its net earnings; but there are some very striking exceptions which do *not* prove the rule. Then some people, acting upon the principle that a road will always pay returns upon its *bonâ fide* cost, say the properties of companies which pay dividends on their stock are worth as much as the balance sheet states. This is undoubtedly true; but there are some lines which cannot be said to be worth less than what they are capitalised at, and yet pay no dividends. So a hard and fast rule does not exist.

With the soundness of a bond it is different. It is quite easy to compile a table showing the rank of various bonds, and by comparing this table with the income accounts of a few years, clear and reliable inferences can be drawn. Let us, as an illustration, take the funded debt statement and the income account of the Louisville and Nashville in its report for 1892. They are given as follows in my larger work on *American Railroads*:—

TABLE I.—FUNDED DEBT OF THE LOUISVILLE AND NASHVILLE RAILROAD COMPANY ON JUNE 30TH, 1892.

Description of Bonds.	Amount.	Matrty.	Rate of interest.	Coupons due	Amt. of interest.
	Dol.		p. c.		Dol.
City of Louisv. Acc. Leb. br. ext.	333,000	1893	6	Apr., Oct.	19,980
*Ten-forty Adjustm	4,531,000	1924	6	May, Nov.	271,860
Louisv. Cinc. & Lex. Ry. mort	2,850,000	1897	7	Jan., July	199,500
Consol. mort. Muinstem and br.	7,070,000	1898	7	Apr., Oct.	494,904
*Memph. & Ohio RR. sterl. mort.	3,500,000	1901	7	June, Dec.	248,780
*Memph, Clarksv. and Louisv. RR. sterl. mort.	2,015,660	1902	6	Feb., Aug.	121,540
Cecelia branch mort	801,000	1907	7	Mar., Sept.	54,600
Louisv. Cinc. & Lex. Ry. 2nd mt.	892,000	1907	7	Apr., Oct.	62,440
Evansv., Henderson & Nashv. div. 1st mort	2,241,000	1919	6	June, Dec.	133,500
Pensacola div. 1st mort.	580,000	1920	6	Mar., Sept	34,800
Southeast & St. L. div. 1st mort.	3,500,000	1921	6	,, ,,	210,000
Pensac. & Atl. RR. 1st mort	1,970,000	1921	6	Feb., Aug.	117,270
*New Orl. & Mobile div. 1st mort.	5,000,000	1930	6	Jan., July	300,000
New Orl. & Mobile div. 2nd mort.	1,000,000	1930	6	,, ,,	60,000
General mortgage bonds	11,458,000	1930	6	June, Dec.	632,860
Louisv. Cinc. & Lex. Ry. gen. mt.	50,000	1931	6	May, Nov.	3,000
First mort, 5 p. c. trust	5,119,000	1931	5	,, ,,	256,450
*First mort. 5 p. c. 56 yr. gold	1,764,000	1937	5	,, ,,	88,200
Southeast, & St. L. div. 2nd mt	3,000,000	1980	3	Mar., Sept.	90,000
*Unified 56 yr. 4 p. c. gold mort.	10,571,000	1940	4	Jan., July	422,840
Ky. Cen. Ry. 1st mort.	6,523,000	1987	4	,, ,,	260,920
Maysv. & Lex. RR. N. div. 7 p. c.	400,000	1906	7	,, ,,	28,000
Maysv. & Lex. RR. S. div. 5 p. c.	219,000	1895	5	June, Dec.	10,950
Total debt	75,397,660			Amount of Interest ...	4,172,394

Condensed Income Account.

Earnings from—	1892—92 Dol.	1890—91 Dol.	1889—90 Dol.	1888—89 Dol.	1887—88 Dol.
Freight	14,604,269	13,113,965	2,845,951	11,325,235	11,081,650
Passengers	5,137,017	4,800,688	4,704,769	4,036,362	4,224,413
Mails	507,136	431,026	422,770	419,050	357,193
Express	455,627	427,425	406,294	368,139	350,811
Gross earnings	21,236,712	19,220,729	18,846,004	16,599,396	16,260,241
Operating expenses—					
Oper. expenses (excl. tax)	13,792,122	12,051,434	11,419,092	10,326,085	10,267,535
Leaves net earnings	7,443,600	7,152,285	7,426,912	6,273,311	6,092,706
Receipts—					
Income from investments	533,293	657,217	638,686	677,109	528,828
Georgia RR. profit	—	60,658	—	—	—
Total income	7,976,893	7,880,160	8,065,598	6,950,420	6,621,534
Disbursements—					
Taxes	513,185	410,810	397,721	401,112	375,557
Rentals				15,000	15,000
Interest on debt, etc.	4,860,878	4,207,629	4,524,694	4,590,493	4,370,803
Georgia RR. deficit	124,695	—	90,339	23,376	3,453
Deficit other RR.	128,001	205,501	186,203	199,425	238,943
Dividends	2,376,000	2,400,000	2,405,367	1,594,800	1,518,000
Total disbursements	8,002,759	7,223,933	7,604,323	6,284,206	6,530,765
Balance	def. 25,866	sur. 656,227	sur. 461,275	sur. 126,214	sur. 90,769

The first of these tables gives details of the various issues whose rank can be easily ascertained after a perusal of the preceding chapter, and this statement further shows that, in 1891-2, the Louisville and Nashville had to pay $4,172,394 interest on its debt. The second says that in the same year it earned $7,976,893, or rather more than $3,800,000 over and above its funded debt; a very considerable margin of safety therefore. The latter of the two tables being " comparative," it will, moreover, be seen that these earnings were nothing out of the common; in fact, they were rather bad, for the company earned less than in preceding years if due allowance is made for the increase of the system; hence it is clear that its bonds are amply secured, and perfectly safe investments. The higher the bonds rank the greater their safety. For example, the first mortgage trust 5 per cent. gold bonds (see Table I) have nothing worth speaking of " in front " of them; the generals, on the other hand, must yield precedence to nearly all other descriptions. Therefore, though still amply secured, they are not quite so good; and if the reader will look at the yield of the two, as given in Appendix A, he will at once see that the "credit" of the first mortgage bonds is better than that of the general bonds. The same table is also of service in determining the value of the shares. By comparing the relation between past earnings and past dividends it becomes possible to draw inferences from current earnings as to dividends in the near future. All that is required to arrive at correct conclusions is great care and due regard for every fact bearing upon the case; the neglect of one single factor of course at once upsets the value of the deductions.

By instituting similar comparisons between the earning capacity of a property, its fixed charges, and the rank of its various securities, even the non-expert is able to satisfy himself as to the safety and intrinsic merit of every security. And it is most desirable that every investor should possess the slight qualifications required for this purpose, because there are many railroad companies which, though of low standing, have nevertheless issued securities of the highest order. To mention a few cases in point, the Reading, Erie and

Northern Pacific are very shaky companies: yet their first mortgage bonds are perfectly safe. This fact, which may strike an outsider as wonderful or incredible, is really only a consequence of those peculiar characteristics of American railroad finance to which we have called attention, an outcome of defects which have provided their own remedy. I entertain no doubt whatsoever that a time will come when the standing of most companies will equal the present position of their best bonds; but the majority of American corporations are still so constituted that they have issued bonds of a far higher status than that which they enjoy themselves. And this circumstance should be fully realised by the investor.

The facts referred to in this little work, and the conclusions deducted from them lead to a very plain inference. The investor must confine himself exclusively to securities whose soundness is beyond a doubt, that is to say securities amply covered by mortgages and earnings, and beyond the reach of the discretion of directors. It matters little whether a bond is issued by a company of the very first order, as long as it is amply secured and as long as earnings leave a sufficient margin of safety above interest requirements. Concerning these facts the investor should satisfy himself. And as there are presumably persons who in spite of the ease with which this can be done, do not care for the pleasure and satisfaction of finding out for themselves, we will in addition to the foregoing remarks give a very simple and fairly reliable rule to go by. The merits of a bond can be inferred from its yield. An American security of the very first order will rarely yield more than £4 10s. per cent. *net*, though up to the £5 level (*net yield*) every bond may be regarded as tolerably safe. It is always a difficult and delicate task to draw the line between the safe and the unsafe. But if it must be drawn somewhere, let it be drawn at this five per cent, because we feel we can answer for that. There are several bonds which yield more, and yet are sound; but those who look upon them as second-class investments will at all events be on the safe side.

With these facts before him the investor ought to find no difficulty in investing his money safely in American securities. A few additional hints might be given, but they are either self-evident, or else a little study of easily obtainable materials will lead to their discovery. For example, it is unwise to purchase bonds when prices are at their highest level. Even the most stable of bonds fluctuates in price. If one looks at the table on page 141, but especially at the Official Weekly List published with the authority of the London Stock Exchange, one will see that all of them reach highest and lowest quotations at pretty regular intervals, and that, on the better class of bonds, the margin between the two levels ranges from about 5 to say 12 per cent. This renders it quite evident that every day in the year is not equally fit for buying bonds. The difference in price makes no very appreciable difference in the net yield. But in addition to the yield there exists an opportunity of making a profit by the regular and recurring appreciation of stocks which, in spite of their oscillations, are sound. By buying and selling *at the proper moment* the investor may therefore add considerably to his income without increasing his risks, or rather, whilst at the same time decreasing both his *risico* and his outlay. Then, it will have been noted that a "guaranteed" bond always yields proportionately more than a company's own security. Thus, Western Pennsylvania fours guaranteed by the Pennsylvania yield £3 19s. 3d. per cent., whereas the Pennsylvania's own equipment fours yield only £3 10s. 9d. Again, the "credit" of Illinois Central three-and-a-halfs is very high, £3 13s. 6d.; that of the Chic. St. Louis and New Orleans fives is £4 6s. 6d., though these bonds are guaranteed by the Illinois Central, and covered by a lien on one of its main lines. The difference made between the two has not much ground to stand upon. The public has simply a prejudice against "guaranteed" bonds; but of course, when the guarantee is good the prejudice is unfounded, and the investor will do well to take advantage of it.

And to return once more to the lesson which it is the principal object of this book to teach : *Shares, other than a few amply covered preferred descriptions like St. Paul or Chicago and Northwestern preferreds, are not fit for investment.* They may be fine things for the gambler, or perhaps are suitable for "speculative investors"; they may be certain of occupying a very high status after a decade or two; they may be "cheap," have "reached rock-bed," or "sure to rise"—but no cautious or conservative investor has any business to trust to them. Perhaps one or two of the very best shares (their names are given in Table IX., on page 172), may be occasionally mixed with bonds. But wise persons had better give a wide berth to all others, no matter whether they pay dividends or not.

In conclusion I give a few tables showing selections of desirable and safe investments of different grades. They were made out before the recent fall occurred, but I see no necessity for alteration as the fall was abnormal, and is in the course of being rectified as this work goes to press. The tables are adapted to the well known principle which spreads a sum of money over various investments in order to reduce risks; and with the help of Appendix A everybody can compile lists that will suit him. Of course, upon a vast number of bells an infinite variety of changes can be rung; but it will be best to select bonds maturing at distant periods, and mixed in such manner that they are secured by liens upon properties in different sections of the country.

List No. 1. First Class Gold Bonds.

		Cost p. ct.	Yield p. ct.
$1,000	Illinois Central 3½ p. ct. due 1951	97	3 13 6
$1,000	Pennsylvania 4 p. ct. Equipmt. Trust due 1914	104	3 17 9
$1,000	New York Central 7 p. ct. 1st Mort. due 1903	126½	4 2 6
$1,000	Cleveland, Cin., Chic. & St. L., Cairo div. 4 p. ct. 1st, due 1939	98	4 4 3
$1,000	Minneapolis, St Sainte Maine & Atlantic 4 p. ct. 1st Mort., due 1926, guar. by Canadian Pacific	96	4 4 9

Investment of £1,043 to yield about £4 0s. per cent.

List No. 2. First Class Sterling Bonds.

		Cost p. ct.	Yield p.ct.
£200	Pennsylvania 6 p. ct. General Mort. due 1910	128	4 0 0
£200	Chicago & Alton, 6 p. ct Consol. Mort. due 1903	118	4 1 9
£200	Baltimore & Ohio 4¼ p. ct. Sterl. Mort. due 1933	112	3 18 3
£200	Philadelphia & Erie 6 p. ct. Sterl. Mort. due 1920	132	4 3 6

Investment of £980 to yield about £4 1s. per cent.

List No 3. First Class Miscellaneous Bonds.

		Cost p. ct.	Yield p ct.
$1,000	Chic., St. Louis, & New Orleans, guar. by Ill. Central, 5 p. ct. Gold, due 1951	122¼	4 2 0
£200	Atlantic First Leased Lines Rent Tr. 4 p. ct. Deb.	94	4 6 6
$1,000	Canada Southern, guar. by N.Y.C., 5 p. ct. 1st Curr., due 1908	105	4 13 9
£200	St. Paul, Minneapolis & Manitoba 4 p.ct. due 1940	92	4 10 6
$1,000	Atchison, Topeka & Sta. Fe, 4 p. ct. General, due 1989	83	4 18 3

Investment of £993 to yield about £4 10s. per cent.

LISTS FOR INVESTORS.

List No. 4. Good Miscellaneous Bonds.

		Cost p. ct.	Yield p. ct.
$1,000	Louisville & Nashville 5 p. ct. 1st Gold due 1931	109	4 10 9
$1,000	Indianapolis & Vincennes 7 p. ct. Curr. due 1908	127½	4 12 6
$1,000	Denver & Rio Grande 4 p. ct. 1st Gold due 1936	89	4 13 6
$1,000	Missouri, Kansas, & Texas 4 p. ct. 1st Gold due 1990	82	4 17 6
$1,000	Oregon & California 5 p. ct. 1st Gold due 1927	99	5 3 0

Investment of £1,013 to yield about £4 15s. per cent.

No. 5. Good Miscellaneous List. Guaranteed Bonds.

		Cost p. ct.	Yield p. ct.
$1,000	Canada Southern 5 p. ct. 1st Currency due 1908, guar. by N. Y. Central ..	105	4 13 9
$1,000	Alleghenny Valley 6 p. ct. Gold, due 1910, guar. by Pennsylvania	128½	4 13 9
$1,000	Nashville, Florence, and Sheffield 5 p. ct. Gold, due 1937, guar. by Louisville and Nashville	105	4 16 9
$1,000	Pittsb., Clevel. and Toledo 6 p. ct. Gold, due 1922, guar. by Balt. and Ohio..	115½	5 1 0
$1,000	Colorado Midland, 6 p. ct. 1st Gold, due 1936, guar. by Atchison	106	5 12 6

Investment of £1,120, to yield about £4 19s. per cent.

No. 6. Fair Miscellaneous List

		Cost p. ct.	Yield p. ct.
$1,000	Norfolk and Western 6 p. ct. General Gold, due 1931....................	122	4 15 9
$1,000	Atchison 4 p. ct. 1st Gen. Gold, due 1989	83	4 18 3
$1,000	Philadelphia and Reading 5 p. ct. Con. Gold, due 1922	100	5 0 0
$1,000	Norfolk and Western 6 p. ct. Improv., due 1934	109	5 10 9
$1,000	N. Y., L. Erie and Western 5 p. ct. Fund. Coup. Gold, due 1969	87½	5 14 3

Investment of £1,003, to yield about £5 5s per cent.

No. 7. Good Second Class Investment.

		Cost p. ct.	Yield p. ct.
$1,000	Pennsylvania $50 shares, 5 p. ct. Consol. Gold, due 1923..................	52¼	5 14 0
£200	Alabama Great Southern 5 p. ct. Genl. Gold, due 1927..................	92	5 11 6
$1,000	Norfolk and Western 5 p. ct. Gold, due 1990	85½	5 18 3
$1,000	Denver and Rio Grande 5 p. ct. Impr. Gold, due 1928..................	85	6 0 6
$1,000	Colorado Midland 4 p. ct. Cons. Gold due 1940	60	6 14 3

Investment of £584, to yield about £6 0s. per cent.

No. 8. Speculative Investment in Bonds.

		Cost p. ct.	Yield p. ct.
$1,000	N. Y., Lake Erie and Western 6 p. ct. 2nd Gold, due 1969	95	6 6 3
$1,000	Chic. and Northern Pacific 5 p. ct. 1st Gold, due 1940..................	78½	6 8 6
$1,000	Colorado Midland 4 p. ct. Consol. Gold, due 1940	60	6 14 3
$2,000	Texas and Pacific 5 p. ct. 1st Gold, due 2000	75	6 13 3
$1,000	East Tenn. Va. and Ga. 5 p. ct. Equip. due 1938	71½	7 8 0

$6,000 nominal.
Investment of £910, to yield about £6 14s. per cent.

9. Fairly Safe Investment in Shares.

		Present Price.	Yield p.ct.	Points which present prices are below highest of year.
$2,000	Baltimore & Ohio Common ..	78x	6 8 3	22⅛
$2,000	Chic., Milw., & St. Paul Preferred	122½	5 15 3	6½
$2,000	Illinois Central Common......	95	5 6 3	12¼
$2,000	Lake Shore Common	125	4 17 6	11½
$2,000	New York Central	104½	4 15 9	10⅞

Investments of £2,100 to yield, *if no change in the rate of Dividend occurs,* about £5 8s. *per cent.*

APPENDIX A.

LIST OF BONDS, ARRANGED ACCORDING TO THEIR YIELD PER CENT., JUNE, 1893.

Amount Outstanding.	Description.	Rate of Int'r'st.	Redeemable.	Price.	Yield Per Cent. £ s. d.	Coupons due.	Remarks.
$9,900,000	Pennsylvania Gold Bonds	4½	1913	114x	3 10 9	July, Dec.	
$2,500,000	Illinois Centl. Gold Mortgage Bonds	3½	1151	97	3 13 6	Jan., July	
£1,800,000	,, ,, Sterling ,, ,,	3½	1950	97	3 13 6	Jan., July	
$11,500,000	New York Central Debentures, Gold	4	1905	103x	3 13 9	July, Dec.	
$15,000,000	Illinois Central 1st Gold Mortgage...	4	1952	106	3 15 9	Apl., Aug.	
$2,000,000	New York Central Sterlg. Mortgage	6	1903	120	3 16 3	Jan., July	
$15,717,000	Pennsylvania Comp. 1st Gold Mort.	4½	1921	112	3 17 0	Jan., July	
$3,000,000	Pennsylvania Equipment Trust Gold Bonds	4	1914	104	3 17 9	Mar., Sept.	
$21,245,000	West Shore 1st Mort. Currency Bnds.	4	—	—	—	—	Guar. by N. Y. Cent.
£713,800	Baltimore and Ohio Sterling ...	6	1895	104	3 17 9	Mar., Sept.	
£2,400,000	Baltimore and Ohio Sterling, 1st ...	4½	1933	112	3 18 3	Apl., Aug.	
$3,000,000	Western Pennsylvania Gold Mort....	4	1928	101x	3 19 3	Apl., Oct.	Guar. by Penns.
£760,900	Chicago, Milwaukee, and St. Paul, Sterling 1st, St. Paul Div. ...	7	1902	129	3 19 3	Jan., July	
$2,781,600	Pennsylvania General Sterling Mort.	6	1910	128	4 0 0	Jan., July	
£200,000	Illinois Central Sterling	5	1905	110½	4 1 6	July, Dec.	Con. into Pref. Stock.
$10,000,000	N. Y. Central Debentures, Currency	5	1901	109	4 1 9	Mar., Sept.	
£875,970	Chicago and Alton Consol. Mortgage	6	1903	118	4 1 9	Jan., July	
$18,000,000	Chicago, St. Louis, and New Orleans Guar. Gold	5	1951	122½	4 2 0	July, Dec.	Guar. by Ill. Central.
$30,000,000	New York Central, 1st Mortgage ...	7	1903	126½	4 2 6	Jan., July	
£5,000,000	Philadelphia and Erie Consolidated	6	1920	132	4 3 6	Jan., July	
$4,650,000	Clevl., Cin., Chicago, and St. Louis 1st Cairo Div.	4	1939	98	4 4 3	Jan., July	

APPENDIX A.—(CONTINUED).

Amount Outstanding.	Description.	Rate of Int'r'st.	Redeemable.	Price.	Yield Per Cent. £ s. d.	Coupons due.	Remarks.
$3,000,000	Pennsylvania Consolidated Sinking Fund Currency	6	1905	119	4 4 6	July, Dec.	
£4,693,800	Pennsylvania Sterling Sinking Fund	6	1905	119	4 4 6	Jan., July	
$8,313,000	Minneapolis S. Ste Marie & Atlantic	4	1926	96	4 4 9	Jan., July	Inter. Guar. by Can. Pacific.
$10,667,000	Baltimore and Ohio South-western	4½	1990	96	4 4 9	Jan., July	Guar. by Balt. & Ohio.
$10,000,000	Balt. and Ohio General Gold Mort.	5	1925	115	4 4 9	Feb., Aug.	
$5,000,000	Long Island Railroads, Gold	5	1931	115	4 4 9	Qtr., Jan.	
£1,920,800	Baltimore and Ohio Sterling, 1st	6	1902	114	4 5 0	Mar., Sept.	
$15,099,000	St. Paul, Minn., and Maint. Gold	4½	1933	106	4 5 9	Jan., July	
$9,500,000	Lehigh Valley, 1st Mortgage, Gold	4½	1940	106	4 6 0	Jan., July	Reduced.
$10,000,000	Balt. and Ohio Consolidated Mort.	5	1988	118	4 6 3	Feb., Aug.	
£800,000	New York and Canada, 1st Sterling	6	1904	115x	4 6 3	May, Nov.	Guar. by Del. and Hudson.
$3,000,000	Baltimore and Potomac Main Line, 1st Gold Mortgage	6	1911	122	4 6 6	Apl., Oct.	Guar. by Penna.
£805,200	Atlantic, 1st Leased Lines Rental Trust Debentures	4	Perp.	94	4 6 6	Jan., July	
£1,399,800	Phil. and Reading General Mort.	6	1911	124	4 7 0	July, Dec.	
$25,340,000	Chicago, Milwaukee, and St. Paul, 1st Gold	5	1921	114	4 7 3	Jan., July	
$3,401,000	Grand Rapids and Indiana (Pennsyl.)	7	1899	117	4 7 9	Jan., July	
$27,229,000	Union Pacific, 1st Mortgage Gold	6	1896-99	110	4 7 9	Jan., July	
£720,000	Illinois Central Sinking Fund	5	1903	105	4 8 0	Apl., Aug.	
£1,353,400	Pittsburgh and Connellsville Cons.	6	1926	130	4 8 0	Jan., July	Guar. by Balt. & Ohio.

APPENDIX A.—(CONTINUED).

Amount Outstanding.	Description.	Rate of Int'r'st.	Redeemable.	Price.	Yield Per Cent. £ s. d.	Coupons due.	Remarks.
$8,000,000	Chicago and N. W. Sink. Fund Curr.	5	1933	112	4 8 0	May, Nov.	
£1,990,600	Baltimore and Ohio Sterling 1st	6	1910	120x	4 9 3	May, Nov.	
$28,545,000	Central of New Jersey, General Gold	5	1987	115	4 9 6	Jan., July	
£2,000,000	St. Paul, Minn. and Manitoba Sterl.	4	1940	92	4 10 0	Jan., July	
$5,129,000	Louisville and Nashville, 1st Gold Mortgage Trust	5	1931	109	4 10 9	May, Nov.	
$3,000,000	Lehigh Valley Consol. Currency	6	1923	124½	4 11 0	July, Dec.	
$24,495,000	Chicago, Burlington, and Quincy Nebraska Ext. Currency	4	—	88	4 11 6	May, Nov.	
$4,000,000	West Virginia and Pittsburg, 1st Gold Mortgage	5	1990	110	4 12 0	Apl., Oct.	Interest guarant'd by Balt. and Ohio.
$1,700,000	Indianapolis & Vincennes, Currency	7	1908	127½	4 12 6	Feb., Aug.	Guar. by Penns. R.R.
$5,600,000	New York, Ontario, and Western, 1st Gold Mortgage	5	1939	107x	4 12 6	July, Dec.	
$28,435,000	Denver and Rio Grande, 1st Gold Mortgage	4	1936	89	4 13 6	Jan., July	
£398,388	Louisville and Nashville (M. C. and L. Div.) 1st	6	1902	110½	4 13 6	Feb., Aug.	Guar. by New York Central until 1988.
$13,923,600	Canada Southern, 1st Currency	5	1908	105	4 13 9	Jun., July	
$1,791,800	Allegheny Valley, 6 p. c. Gold, or 7 p. c. Currency	7	1910	128¾	4 13 9	Apl., Oct.	Guar. by Penns. Railway.
$1,300,000	Chicago, Milwaukee, and St. Paul (Wis. and Minn. Div.) Gold	5	1921	107	4 13 9	Jan., July	
$16,980,000	New York, Lake Erie and Western 1st Gold	7	1920	137½	4 14 0	Mar., Sept.	

APPENDIX A.—(CONTINUED).

Amount Outstanding.	Description.	Rate of Int'r'st.	Redeemable.	Price.	Yield Per Cent. £ s. d.	Coupons due.	Remarks.
$6,500,000	Burlington, Cedar Rapids and Northern, 1st Mortgage Currency	5	1906	105½	4 14 9	July, Dec.	
$7,750,000	Louisville and Nashville Unified Gold	4	1940	85	4 14 9	Jan., July	
$31,361,500	Southern Pacific of California, 1st Gold Mortgage	6	1905-12	116½	4 15 0	Apl., Aug.	
£948,200	South and North Alabama, Gold 1st	6	1903	110x	4 15 0	May, Nov.	Louisvl. and Nashvl.
£700,000	Louisville and Nashville (Ohio and Miss. Div.) Gold	7	1901	118½	4 15 6	July, Dec.	
$2,204,000	Chesapeake and Ohio Consolidated, 1st Mortgage Gold	5	1939	104	4 15 6	May, Nov.	
$7,283,000	Norfolk and Western General Gold Mortgage	6	1931	122	4 15 9	May, Nov.	
$1,750,000	Alabama, Great Southern, 1st Gold Mortgage	6	1908	114½	4 15 9	Jan., July	[fore 1921.
$14,263,000	Northern Pacific, 1st Gold Gen.	6	1921	120	4 16 6	Jan., July	Drawings at 110 be-
$2,059,000	Nashville, Florence & Sheffield Gold	5	1937	105	4 16 9	Feb., Aug.	Louisvl. and Nashvl.
$1,960,000	Louis. and Nash. Sink. Fund, Gold	6	1910	111¼	4 17 0	Apl., Oct.	
$39,466,000	Missouri, Kansas and Texas, 1st Gold	4	1990	82	4 17 6	July, Dec.	
$2,049,000	Chicago, Milwaukee, and St. Paul (Chic. and Mo. Div.)	5	1926	104	4 17 9	Jan., July	
$5,360,000	Georgia, Carolina, and Northern	5	1929	104	4 18 0	Jan., July	
$129,493,000	Atchison, Topeka, and Santa Fé, 1st Gen. Gold Mortgage	4	1989	83	4 18 3	Jan., July	
$31,997,000	Chic., Rock Isl. and Pac., 1st Curr.	5	1934	103¾	4 18 6	Jan., July	
$20,000,000	Louisville and Nashville Gen. Gold	6	1930	118x	4 18 9	July, Dec.	

APPENDIX A.—(CONTINUED).

Amount Outstanding.	Description.	Rate of Int'r'st.	Redeemable.	Price.	Yield Per Cent. £ s. d.	Coupons due.	Remarks.
$1,500,000	Wheeling and Lake Erie (Wheeling Div.) 1st Gold	5	1928	103	4 19 0	Jan., July	
$3,705,977	New York, Lake Erie, and West, 1st Gold Mortgage, Fund, Coup.	7	1920	132½	4 19 3	Mar., Sept.	
$5,768,452	Philadelphia and Read. Consol. Gold	5	1922	100x	5 0 0	May, Nov.	
$1,872,800	Phil. and Reading Improv. Mort.	6	1897	104	5 0 3	Apl., Aug.	
$2,400,000	Pittsburg, Clevel., and Toledo Gold	6	1922	115½	5 1 0	Apl., Oct.	Guar. by Balt. & Ohio.
$1,071,000	St. Louis and San Francisco (Mi. and W. Div.) 1st Guar.	6	1919	115	5 1 0	Feb., Aug.	
$1,400,000	Wheeling and Lake Erie, Gold	5	1930	110	5 2 0	Feb., Aug.	
$16,654,000	Oregon and California, 1st Gold	5	1927	99	5 3 0	Jan., July	Guar. by S. Pacific.
$5,680,000	Chicago, Milwaukee, and St. Paul Hat. and Dak. Div.) 1st Curr.	7	1910	123¾	5 3 9	Jan., July	
$14,000,000	Rio Grande Western, 1st Gold	4	1939	78	5 3 0	Jan., July	
$10,500,000	Chic. and Western Indiana Gen. Gold	6	1932	113½	5 4 0	Qtr., Mar.	
£367,000	Northern Central Consolidated	6	1904	108	5 4 3	Jan., July	
$5,000,000	St. Louis Bridge, 1st Mort, Gold	7	1929	130	5 4 9	Apl., Oct.	
$1,750,000	Rio Grande Junction Gold	5	1939	94	5 7 0	July, Dec.	
$35,703,000	Philadelphia and Reading Gen. Gold	4	1958	75	5 8 0	Jan., July	
£748,850	Alabama, New Orleans, and Texas Pacific Junction ("A." Deb.)	5	—	92x	5 8 9	May, Nov.	
$1,125,000	Perkiomen, 2nd series, Gold	5	1918	94½	5 9 0	Qtr., Jan.	Guar. by Reading.
$5,000,000	Norfolk and Western Improvement	6	1934	109	5 10 9	Feb., Aug.	
£476,000	Alabama Gt. South. Gen. Gold Mort.	5	1927	92	5 11 6	July, Dec.	
$6,250,000	Colorado Midland, 1st Mort. Gold	6	1936	106x	5 12 6	July, Dec.	Atchison System.

APPENDIX A.—(CONCLUDED).

Amount Outstanding.	Description.	Rate of Int'r'st.	Redeemable.	Price.	Yield Per Cent. £ s. d.	Coupons due.	Remarks.
$1,323,000	Vicksburg, Shreveport and Pacific, Prior Lien	6	1915	106	5 13 9	May, Nov.	
$4,029,840	N.Y., L. Erie & W. Fund Coup. Gold	5	1969	87½x	5 14 3	July, Dec	
$7,200,000	Norfolk and Western, 100-yr. Gold	5	1990	85½	5 18 3	Jan., July	
$2,800,000	Alabama Midland, 1st Gold	6	1928	92	6 0 0	May, Nov.	
$8,050,000	Denver & Rio Gra. Imp. Mort. Gold	5	1928	85	6 0 6	July, Dec.	
$4,500,000	Galveston and Harrisburg Gold	6	1910	111½	6 2 3	Feb., Aug.	
$5,500,000	Oregon Short Line and Utah North. Coll. Tr.	5	1919	81½	6 4 3	Mar., Sept.	Union Pacific System.
$33,597,400	N.Y., L. Erie, & West., 2nd Gold Mort.	5	1919	81½	6 4 3	Mar., Sept.	
$1,500,000	Norfolk and Western Adjustment Gold Mortgage	6	1969	95x	6 6 3	July, Dec.	
$7,000,000	St. Joseph & Gr. Isl., 1st Gold Mort.	7	1924	109x	6 7 0	Qtr., Mar.	
$24,915,000	Chicago and North Pacific, 1st Gold Mortgage	6	1925	94½	6 8 6	May, Nov.	
$2,090,000	Chattanooga, Rome, and Col., 1st Gold Mort.	5	1940	78½	6 8 6	Apl., Oct.	North. Pacific System
$21,049,000	Texas and Pacific, 1st Mort., Gold	5	1937	80	6 9 0	Mar., Sept.	
$1,809,000	Colorado Midland, Cons. Mort. Gold	5	2000	75x	6 13 3	July, Dec.	
$5,631,000	Northern Pac. and Mont., 1st Gold	4	1940	60	6 14 3	Feb., Aug.	
$1,348,000	Richmond and Danville, 1st Mort.	6	1938	88	6 18 6	Mar., Sept.	
$4,500,000	East Tenn., Va., and Ga. Equipment Mortgage	5	1909	82	7 0 6	Mar., Sept.	Plant System.
$40,930,000	Northern Pacific Land Grant	5	1938	71½	7 8 0	Mar., Sept.	
		5	1989	67½	7 8 3	July, Dec.	
$3,000,000	Mobile and Birmingham, 1st Gold	5	1937	47½	nil.	Jan., July	Guar. E.T., Va. & Ga.

APPENDIX B.

Table showing Dividend paying Shares and Income Bonds, with rate of Dividend paid during the past Six Years.

Amount Outstanding	Description	Face Value	Dividend (p. ct. per annum) paid.						Remarks
			1892	1891	1890	1889	1888	1887	
£676,070	Alabama, Gt. South. 6 p.c. Pref. ("A")	£10	6	6	9	6	—	—	
£1,566,000	Do. Ordinary or "B"	"	1½	1½	4	4	—	—	
£805,500	Atlantic 1st Leased Lines Rental Tr.	"	4½	4½	—	—	4	4	
$30,000,000	Baltimore and Ohio, Common	$100	5	—	6	3	—	4	20 p. ct. stock div. in 1891.
$22,488,000	Central Railroad of New Jersey	"	7	6½	6	2	2	2	
$100,000,000	Central Pacific of California	"	2	2	2	2	2½	5	(Guar. by South Pacific.
$16,027,261	Chicago, Milwaukee & St. Paul, Com.	"	4	—	—	—	7	6	
$25,767,900	Do. Non-cumul. Pref.	"	7	7	7	7	7	7	
$39,054,383	Chicago & North-Western, Common	"	6	6	6	6	6	6	
$22,335,100	Do. Pref. Non-cum.	"	7	7	7	7	7	7	
$11,236,850	Cleveland and Pittsburg	$50	7	7	7	7	7	7	Guar. by Penna. R.R.
$28,000,000	Cleveland, Cincinnati, Chicago and St. Louis, Common	$100	3	3	4½	—	—	—	Company dates from 1890.
$10,000,000	Do. Non-cumulative Pref.	"	5	5	5	7	6	5	
$30,000,000	Delaware and Hudson Canal	"	7	7	7	2¼	3¼	2½	1¼ of 1888 div. was in scrip afterwards redeemed, paid 1 p.ct. twice in 1893, but passed July dividend.
$28,000,000	Denver and Rio Grande, Preferred	"	—	—	2¼				
$50,000,000	Illinois Central, Common	"	5	5	6	5½	7	7	
$10,000,000	Do. Leased Line, 4 p.c. Stock	"	4	4	4	4	4	4	

APPENDIX B.—(CONCLUDED).

Amount Outstanding.	Description.	Face Value.	Dividend (p. ct. per annum) paid.							Remarks.
			1892.	1891.	1890.	1889.	1888.	1887.		
$49,465,500	Lake Shore and Michigan Southern	$100	6¼	6¼	5	5	4	4		100 p. ct. stock div. in 1880; 4.9 p.c. of 1890 div. was stock. Stk. to be incrsd. to $100,000,000 Div. discontinued again.
$52,800,000	Louisville and Nashville	"	4½	5	6	5	5	—		
$89,428,300	New York Central and Hudson River	"	5¼	4½	4½	4	4	4		
$8,536,600	New York, L. Erie & Western, Pref.	"	3	—	—	—	—	—		
$8,000,000	New York, Susquehanna, and West. Cumulative Pref.	"	3	1¼	—	—	—	—		Cumul. div. due = 55 p. ct. 1 p.c. of 1892 div. was in scrip, div. April, 1893, was passed.
$43,000,000	Norfolk and Western 6 p. ct. Non-cumulative, Preferred	"	2½	3	3	3	1½	—		Guar. by Reading. Discontinued since Ap., 1892. 1893, 1st half, paid 2½ cash and 2 stock.
$1,720,750	North Pennsylvania	$50	8	8	8	8	8	8		
$36,288,750	Northern Pacific 8 p.c. Non-cum. Pref.	$100	2	4	4	—	—	—		
$126,774,500	Pennsylvania	$50	6	6	5½	5	5	5½		
$34,084,575	Pittsburg, Fort Wayne and Chicago (guar. 7 p. ct. by Penna Railroad)	$100	7	7	7	7	7	7		
$7,500,000	Rio Grande Western, Pref. Stock	"	5	5	—	—	—	—		
$2,490,300	St. Louis Bridge, 1st Pref. } jointly guar.	"	6	6	6	6	6	6		
$3,000,000	Do. 2nd " } by Wa. and Mis. Pacific.	"	3	0	0	0	0	0		
$1,250,000	Tunnel Railway of St. Louis	"	6	6	6	6	6	6		
$20,000,000	St. Paul, Minneapolis and Manitoba	"	6	6	6	6	6	6		Guar. by Great Northern.
$4,500,000	Wheeling and Lake Erie, 6 per cent. Non-cumulative, Preferred	"	4¾	4¾	4	4	3	—		

APPENDIX C.

List of Non-Dividend paying Shares, quoted in London, with highest and lowest prices for 1893, and date of last Dividend, if the Stock has ever received any.

Amount Outstanding.	Description.	Par Value.	Highest and Lowest London Quotations. January 1st to June 15th, 1893.	
			Highest.	Lowest.
£1,500,000	Alab., New Orleans and Texas Pacific, Pref. "A."	£10	1	$\frac{3}{4}$
£2,500,000	Do. Do. Def. "B."	,,	$\frac{3}{8}$	$\frac{1}{2}$
$102,000,000	* Atchison, Topeka, and Santa Fé	$100	$37\frac{3}{8}$	$23\frac{9}{16}$
$60,187,100	Chesapeake and Ohio, Common	,,	26	19
$38,000,000	Denver and Rio Grande, Common	,,	19	$13\frac{1}{2}$
$27,500,000	East Tennessee, Virginia, and Tenn., Common ..	,,	—	—
$18,500,000	Do. Do. 2nd Preferred ..	,,	12	6
$17,000,000	Missouri, Kansas and Texas, Common	,,	$16\frac{3}{8}$	11
$13,000,000	Do. Do. (New) Preferred ..	,,	$27\frac{1}{4}$	nom.
$77,427,000	New York, Lake Erie and Western, Common ..	,,	$27\frac{3}{4}$	$16\frac{7}{8}$
$58,113,982	New York, Ontario and Western, Common ..	,,	$20\frac{1}{16}$	$14\frac{1}{2}$
$9,500,000	Norfolk and Western, Common	,,	—	—
$49,000,000	Northern Pacific, Common	,,	$17\frac{1}{16}$	$13\frac{1}{4}$
$39,830,362	Philadelphia and Reading, Common	$50	$27\frac{3}{8}$	$8\frac{1}{4}$
$16,509,000	St. Louis, South-Western, Common..	,,	—	—
$20,000,000	Do. 5 per cent. Non-cumulative Preferred ..	,,	12	nom.
$60,868,500	† Union Pacific	,,	$43\frac{3}{4}$	$27\frac{1}{4}$
$28,000,000	Wabash, Common	,,	$12\frac{3}{8}$	$8\frac{1}{2}$
$24,000,000	Wabash, Preferred	,,	$26\frac{3}{4}$	$15\frac{3}{4}$

* Last Dividend paid Nov. 15th, 1888. † Last Dividend paid April 1st, 1884.

APPENDIX D.—TABLE FOR INVESTORS.

The following Tables show the rate per cent. of annual income realized from shares bearing any given rate of yearly dividends or interest, from 1 to 15 per cent., when purchased at various prices from 10 to 200 per cent.

For example: To ascertain what rate of annual interest will be realized on a share which bears 7 per cent. per annum, and can be purchased at 92 (*i.e.*, at 92 per cent. of its par value, whatever the par may be), find 92 in the column of "purchase price," and follow that line across to the column headed "7 per cent.," which will show the correct figure—in the present instance, 7.60 per cent.

Purchase Price.	3 per cent.	4 per cent.	4½ per cent.	5 per cent.	6 per cent.	7 per cent.	8 per cent.	9 per cent.	10 per cent.	12 per cent.	15 per cent.
10	30	40	45	50	60	70	80	90	100	120	150
15	20	26.66	30	33.33	40	46.66	53.33	60	66.66	80	100
20	15	20	22.50	25	30	35	40	45	50	60	75
22	13.63	18.18	20.45	22.72	27.27	31.81	36.36	40.90	45.45	54.54	68.18
24	12.50	16.66	18.75	20.83	25	29.16	33.33	37.50	41.66	50	62.50
26	11.50	15.38	17.30	19.33	23.07	26.92	30.76	34.61	38.46	46.15	57.69
28	10.71	14.28	16.07	17.85	21.42	25	28.57	32.14	35.71	42.85	53.57
30	10	13.33	15	16.66	20	23.33	26.66	30	33.33	40	50
32	9.37	12.50	14.06	15.82	18.75	21.87	25	28.12	31.25	37.50	46.87
34	8.82	11.76	13.23	14.70	17.64	20.58	23.52	26.47	29.41	35.29	41.11
36	8.33	11.11	12.50	13.88	16.66	19.44	22.22	25	27.77	33.33	41.66
38	7.89	10.52	11.84	13.15	15.78	18.42	21.50	23.68	26.31	31.57	39.47
40	7.50	10	11.25	12.50	15	17.50	20	22.50	25	30	37.50
42	7.14	9.52	10.71	11.90	14.28	16.66	19.04	21.52	23.80	28.57	35.71
44	6.81	9.09	10.22	11.36	13.63	15.90	18.18	20.45	22.72	27.27	34.09

APPENDIX D.—(CONTINUED).

Purchase Price	3 per cent.	4 per cent.	4½ per cent.	5 per cent.	6 per cent.	7 per cent.	8 per cent.	9 per cent.	10 per cent.	12 per cent.	15 per cent.
46	6.52	8.69	9.78	10.86	13.04	15.21	17.39	19.56	21.73	26.08	32.60
48	6.25	8.33	9.37	10.41	12.50	14.58	16.66	18.75	20.83	25	31.25
50	6	8	9	10	12	14	16	18	20	24	30
51	5.88	7.84	8.82	9.80	11.76	13.72	15.68	17.64	19.60	23.52	29.41
52	5.76	7.69	8.65	9.61	11.53	13.46	15.38	17.30	19.23	23.07	28.84
53	5.66	7.54	8.49	9.43	11.32	13.20	15.09	16.98	18.86	22.64	28.30
54	5.55	7.40	8.33	9.25	11.11	12.96	14.81	16.66	18.51	22.22	27.77
55	5.45	7.27	8.18	9.09	10.90	12.72	14.54	16.36	8.18	21.81	27.27
56	5.35	7.14	8.03	8.92	10.70	12.50	14.28	16.07	17.85	21.42	26.78
57	5.26	7.01	7.89	8.77	10.52	12.27	14.03	15.78	17.54	21.05	26.31
58	5.17	6.89	7.75	8.62	10.34	12.06	13.79	15.51	17.24	20.68	25.86
59	5.08	6.77	7.62	8.47	10.16	11.86	13.55	15.25	16.94	20.33	25.42
60	5	6.66	7.50	8.33	10	11.66	13.33	15	16.66	20	25
61	4.91	6.55	7.37	8.19	9.83	11.47	13.11	14.75	16.39	19.67	24.59
62	4.83	6.45	7.25	8.06	9.67	11.29	12.90	14.51	16.12	19.35	24.19
63	4.76	6.34	7.14	7.93	9.52	11.11	12.69	14.28	15.87	19.04	23.80
64	4.68	6.25	7.03	7.81	9.37	10.93	12.50	14.06	15.62	18.75	23.43
65	4.61	6.15	6.92	7.69	9.23	10.76	12.30	13.84	15.38	18.46	23.07
66	4.54	6.06	6.81	7.57	9.09	10.60	12.12	13.63	15.15	18.18	22.72
67	4.47	5.97	6.71	7.46	8.95	10.44	11.94	13.43	14.92	17.91	22.38
68	4.41	5.88	6.61	7.35	8.82	10.29	11.76	13.23	14.70	17.64	22.05
69	4.34	5.79	6.52	7.24	8.69	10.14	11.59	13.04	14.49	17.39	21.73
70	4.28	5.71	6.42	7.14	8.57	10	11.43	12.85	14.28	17.40	21.42
71	4.22	5.63	6.33	7.04	8.45	9.85	11.26	12.67	14.08	16.90	21.12
72	4.16	5.55	6.25	6.94	8.33	9.72	11.11	12.50	13.89	16.66	20.83

APPENDIX D.—(CONTINUED).

Purchase Price.	3 per cent.	4 per cent.	4½ per cent.	5 per cent.	6 per cent.	7 per cent.	8 per cent.	9 per cent.	10 per cent.	12 per cent.	15 per cent.
73	4.10	5.47	6.16	6.84	8.21	9.58	10.95	12.32	13.69	16.43	20.54
74	4.05	5.40	6.08	6.75	8.10	9.45	10.80	12.16	13.51	16.21	20.27
75	4	5.33	6	6.66	8	9.33	10.66	12	13.33	16	20
76	3.94	5.26	5.92	6.57	7.89	9.21	10.52	11.84	13.15	15.78	19.73
77	3.89	5.19	5.84	6.49	7.79	9.09	10.38	11.68	12.98	15.58	19.48
78	3.84	5.12	5.76	6.41	7.69	8.97	10.25	11.53	12.82	15.38	19.23
79	3.79	5.06	5.69	6.32	7.59	8.86	10.12	11.39	12.65	15.18	18.98
80	3.75	5	5.62	6.25	7.50	8.75	10	11.25	12.50	15	18.75
81	3.70	4.93	5.55	6.17	7.40	8.64	9.87	11.11	12.34	14.81	18.51
82	3.65	4.87	5.48	6.09	7.31	8.53	9.75	10.97	12.19	14.63	18.29
83	3.61	4.81	5.42	6.02	7.22	8.43	9.63	10.84	12.04	14.45	18.04
84	3.57	4.76	5.35	5.95	7.14	8.33	9.52	10.71	11.90	14.28	17.85
85	3.52	4.70	5.29	5.88	7.05	8.23	9.41	10.58	11.76	14.11	17.61
86	3.48	4.65	5.23	5.81	6.97	8.13	9.30	10.46	11.62	13.95	17.44
87	3.44	4.59	5.17	5.74	6.89	8.04	9.19	10.34	11.49	13.79	17.24
88	3.40	4.54	5.11	5.68	6.81	7.94	9.09	10.22	11.36	13.63	17.04
89	3.37	4.49	5.05	5.61	6.74	7.86	8.98	10.11	11.23	13.48	16.85
90	3.33	4.44	5	5.55	6.66	7.77	8.88	10	11.11	13.33	16.66
91	3.29	4.39	4.94	5.49	6.59	7.69	8.79	9.89	10.98	13.18	16.48
92	3.26	4.34	4.89	5.43	6.52	7.60	8.69	9.78	10.86	13.04	16.30
93	3.22	4.30	4.83	5.37	6.46	7.52	8.60	9.67	10.75	12.90	16.12
94	3.19	4.25	4.78	5.31	6.38	7.44	8.51	9.57	10.63	12.76	15.95
95	3.15	4.21	4.73	5.26	6.31	7.36	8.42	9.47	10.52	12.63	15.78

OCTOBER, 1899.

CATALOGUE

OF

Legal, Commercial

and other Works

PUBLISHED AND SOLD BY

EFFINGHAM WILSON

Publisher and Bookseller.

11 ROYAL EXCHANGE, LONDON.

TO WHICH IS ADDED A LIST OF

TELEGRAPH CODES.

EFFINGHAM WILSON undertakes the printing
and publishing of Pamphlets and Books of every
description upon Commission. Estimates given, and
conditions of Publication may be had on application

INDEX.

	PAGE		PAGE
Arbitrage—		**Clerks** (*continued*)—	
Haupt, O. (Arbitrages et Parités)	17	Merchant's	
Willdey's American Stocks	26	School to Office	
Arbitration—		Solicitor's	1
London Chamber of	24	**Correspondence (Commercial)—**	
Lynch, H. Foulks	20	Beaure	1
Banking—		Martin (Stockbrokers)	
Banking, History of	10	Coumbe	1
Banks and their Customers	5	**Counting-house—**	
Banks, Bankers and Banking	22	Crowley	1
Bibliography (Bank of England)	25	Pearce	
Easton's Banks and Banking	15	Tate	2
Easton's Work of a Bank	15		
English and Foreign (Attfield)	11	**County Court—**	
Examination Questions, Arithmetic and Algebra	21	Jones	1
Half-yearly Balance Sheets	11	**Currency and Finance—**	
Howarth's Clearing Houses	18	Aldenham (Lord)	1
Hutchison, J.	18	Barclay (Robert)	1
Journal Institute of Bankers	19	Clare's Money Market Primer	1
Questions on Banking Practice	23	Cobb	1
Scottish Banking	19	Cuthbertson	1
Smith's Banker and Customer	24	Del Mar's History	1
Bankruptcy—		Del Mar's Science of Money	1
McEwen (Accounts)	20	Ellis	1
Stewart (Law of)	7	Gibbs, Hon. H., Bimetallic Primer	1
Bills of Exchange—		Haupt	1
Kölkenbeck (Stamp Duties on)	19	Indian Coinage and Currency	2
Smith (Law of Bills, etc.)	7	Poor (H. V.) The Money Question	2
Bimetallism—		**Dictionaries—**	
List of Works	27, 28	Méliot's French and English	2
Book-keeping—		**Directors—**	
Cariss	13	Haycraft (Liabilities and Duties)	1
Carr (Investors)	11	**Exchanges—**	
Drapers' Accounts	15	Brazilian Exchanges	2
Harlow's Examination Questions	17	Clare	1
Holah's Double Entry	10	Goschen	1
Jackson	18	Norman's Universal Cambist	2
Richardson's Weekly Newspapers	23	Tate's Modern Cambist	2
Sawyer	24	**Exchange Tables—**	
Seebohm's (Theory)	10	Dollar (Eastern)	1
Sheffield (Solicitors)	24	Garratt (South American)	1
Van de Linde	25	Goodricke's Tea Exchange Tables	1
Warner (Stock Exchange)	26	Lecoffre (French)	2
Clerks—		,, (Austria and Holland)	2
Commercial Handbook	10	Merces (Indian)	2
Companion to "Solicitor's Clerk"	19	Schultz (American)	2
Corn Trade	23	Schultz (German)	2
Counting-house Guide	25	**Insurance—**	
Kennedy (Stockbrokers)	8	Bourne's Publications	1
Mercantile Practice (Johnson)	19	Short-Term Table	2

	PAGE		PAGE
nterest Tables—		**Miscellaneous** (*continued*)—	
Bosanquet	12	On Compound Interest and Annuities	24
Crosbie and Law (Product)	13		
Cummins (2¼ %) . .	13	Copyright Law	13
Gilbert's Interest and Contango	16	Cotton Trade of Great Britain .	15
Gunnersall	17	District and Parish Councils (Lithiby)	20
Ham (Panton) Universal .	17		
Indian Interest (Merces) .	22	Factors (Law relating to) .	11
Lewis (Time Tables) . .	20	Gresham, Sir Thomas (Life of) .	12
Rutter	24	Ham's Customs Year Book .	17
Schultz	24	Ham's Inland Revenue Year Book	17
Wilhelm (Compound) . .	26	High Court Practice . . .	23
nvestors (*see* also Stock Exchange Manuals)—		Licensing Acts	19
		Macfee, K. N., Imperial Customs Union	20
Birk's Investment Ledger .	11		
Investment Profit Tables .	27	Maritime Codes, Spain and Portugal (Raikes) . . .	23
Houses and Land . .	9		
How Money makes Money .	15	Maritime Codes, Holland and Belgium	23
How to Invest Money .	9		
oint-Stock Companies—		Patent Law and Practice (Emery)	15
Cummins' Formation of Accounts	13	Property Law (Maude) . .	21
Company Frauds Abolition .	5	Public Man	26
Company Promoters (Law of) .	5	Public Meetings . . .	26
Haycraft (Directors) . .	10	Red Palmer	25
How to ascertain the Profits (Dale)	14	Schedule D of Income Tax .	10
Simonson's Debentures and Debenture Stock (Law of) . .	24	Solicitors' Forms (Charles Jones)	19
		Veld and "Street" . . .	17
Smith	7	World's Statistics . . .	11
egal and Useful Handy Books—		**Money Market** (*see* Currency and Finance).	
List of	7-10	**Pamphlets**	27
aps—		**Prices—**	
British Columbia . .	6	Dunsford (Railways) . .	15
Hauraki Goldfields . .	6	Ellis (Market Fluctuations) .	15
Kalgoorlie	6	Mathieson (Stocks) . . .	21
Tasmania, West Coast of .	6	**Railways—**	
Witwatersrand Goldfields .	6	American and British Investors .	26
ining—		Dunsford (Dividends and Prices)	15
Accounts of G. M. Cos. .	14	Home Rails as Investments .	25
Beeman's Australian Mining Manual	11	Mathieson's Traffics . .	21
		Poor's Manual (American) .	23
British Columbia Mining Laws	12	Railroad Report (Anatomy of a)	27
Charlton's Information for Gold Mining Investors . .	13	Railways in India . . .	22
		Ready Reckoners (*see* also Exchange Tables, Interest, etc.)—	
Gabbott's How to Invest in Mines	16		
Goldmann (South African Mining)	16	Buyers and Sellers' (Ferguson) .	9
Kindell's African Market Manual	19	Commission and Brokerage .	22
Milford's Dictionary of Mining Terms	22	Hawke's Instantaneous Share Reckoner	17
		Henselin's (Multiplication) .	18
Paull's Columbia and Klondyke Manual	22	Ingram (Yards) . . .	18
		Kilogramme Tables . . .	25
Tin-Mining in Spain . .	12	Redeemable Stocks (Mathieson)	21
iscellaneous—		Merces (Indian) . . .	22
Arithmetic and Algebra .	21	Robinson (Share) . . .	23
Australia in 1897 . .	22	Silver Tables (Bar Silver) . .	16
Author's Guide . . .	27		

	PAGE		PAGE
Sinking Fund and Annuity Tables—		**Stock Exchange Manuals,etc.**(*cont.*)—	
Booth and Grainger (Diagram)	12	Robinson (Share Tables)	25
Hughes	18	Rules and Usages (Stutfield)	25
Speculation (*see* Investors and Stock Exchange).		Stock Exchange Official Intelligence	25
		Stock Exchange Values, 1885-1895	26
Stock Exchange Manuals, etc.—		Willdey's American Stocks	26
Contango Tables	16	**Tables** (*see* Exchange Tables, Interest Tables, Ready Reckoners, and Sinking Fund and Annuity Tables, etc.).	
Fenn on the Funds, English and Foreign	15		
Higgins, Leonard, The Put-and-Call	18	**Telegraph Codes—**	
Investor's Ledger	21	Ager's (list of)	29, 30
Investors' Tables, Permanent or Redeemable Stocks	18	Miscellaneous (list of)	30, 31
		The Premier Code	32
Laws and Customs (Melsheimer)	21	**Trustees—**	
Laws, English and Foreign Funds (Royle)	24	Investment of Trust Funds	7
Motors and Cycles	18	Judicial Trustees Act, 1896	19
Options (Castelli)	13	Marrack's Statutory Trust Investments	20
Poor's American Railroad Manual	23		
Rapid Share Calculator	14	Trustees, their Duties, etc.	
Redeemable Stocks (a Diagram)	12	**Wilson's Legal and Useful Handy Books List**	7-10
Registration of Transfers	15		

NEW BOOKS.

COMPANY FRAUDS' ABOLITION.

Suggested by

A review of the Company Law for more than half a century.

By

RICHARD RUSSELL.

Price 1s. 6d.

Just Published, Third Edition, Price 1s. net.

BANKS AND THEIR CUSTOMERS.

A Practical Guide for all who keep Banking Accounts from the Customer's point of view.

BY

The Author of "The Banks and the Public".

MAPS.

NEW MAP OF THE WITWATERSRAND GOLD-FIELDS. Compiled by Messrs. WOOD and ORTLEPP of Johannesburg. Scale, half-mile to the inch. Size, 9 feet by 3 feet. Prices: four Coloured Sheets, £4 4s.; Mounted to fold in Case, £5 14s. 6d.; Rollers Varnished, £5 14s. 6d.; Mounted in Portfolio, £6 6s.; Mounted on Spring Rollers, £12 12s.

KALGOORLIE. Showing the Gold Mining Leases in the direct Hannan's Belt, East Coolgardie Goldfield, Western Australia. Price on Roller and Varnished, 15s. net. Mounted to fold, in Case, 21s. net.

MAPS—*continued*.

HANNAN'S GOLD FIELDS, WEST AUSTRALIA.
An entirely New Map. Scale, 10 chains to the inch. Size, 9 feet by 3 feet. Showing the Lodes and Boundaries, Pipe Lines, Shafts, Batteries, with number of Stamps, etc. Prices: three Coloured Sheets, £3 3s.; Mounted to fold in Case, £4 14s. 6d.; Mounted on Rollers and Varnished, £4 14s. 6d.

A NEW MAP OF THE BOULDER GROUP OF THE HANNAN'S GOLD FIELD, KALGOORLIE.
(The famous "Australia Square Mile".) Scale, 20 inches to one mile. Size, 40 inches by 30 inches. Price 20s. net, folded in Case or Mounted on Rollers.

NEW MAP OF THE WEST COAST OF TASMANIA.
Showing the General Features of the Country, Railways, Harbours, and principal Mineral Fields. Compiled by Lieut.-Colonel BODDAM, late Commanding Engineer, Tasmania. Price 6s.

HAURAKI GOLDFIELDS: New Zealand, Geology and Veins.
With Coloured Maps and Plates in separate cover. By JAMES PARK, F.G.S. Issued by the New Zealand Institute of Mining Engineers. Price 10s. net.

BRITISH COLUMBIA.

THE "PROVINCE" MINING MAPS.

Cariboo. Map of the Central District. Price 4s. net.

West Kootenay Central Division, 4 Maps. Price 4s. 6d. net.

West Kootenay Southern Division, 4 Maps. Price 4s. 6d. net.

Klondike and the Canadian Yukon, and Routes thereto, from the latest official sources. Sheet. Price 2s.

WILSON'S
LEGAL AND USEFUL HANDY BOOKS.

PRICES ALL NET.

The Law of Residential and Business Flats.
By GEO. BLACKWELL, Esq., of the Inner Temple, Barrister-at-Law. Price 1s. 6d.

Law of Bills, Cheques, Notes and I O U's.
Sixty-first Thousand. By JAMES WALTER SMITH, Esq., LL.D., of the Inner Temple, Barrister-at-Law. Price 1s. 6d.

Joint-Stock Companies (1862-1890).
New and Revised Edition. Twenty-fifth Thousand. By JAMES WALTER SMITH, Esq., LL.D. Price 1s. 6d.

The Law of Private Trading Partnership (including the 1890 Act).
Twenty-eighth Thousand. New and Revised Edition. By JAMES WALTER SMITH, Esq., LL.D. Price 1s. 6d.

Master and Servant. Employer and Employed, including the "Workmen's Compensation Act, 1897".
Seventeenth Thousand. By JAMES WALTER SMITH, Esq., LL.D. Price 1s. 6d.

Husband and Wife.
Engagements to Marry, Divorce and Separation, Children, etc. By JAMES WALTER SMITH, Esq., LL.D. Eleventh Thousand. Price 2s. 6d.

Owner, Builder and Architect.
By JAMES WALTER SMITH. Price 1s.

Law of Trustees under the Act, 1893, and the Judicial Trustees Act of 1896.
Their Duties and Liabilities. New and Revised Edition. By R. DENNY URLIN, Esq., of the Middle Temple, Barrister-at-Law. Price 1s.

The Investment of Trust Funds under the Trustee Act, 1893.
By R. DENNY URLIN, Esq. Price 1s.

Law of Wills.
A Practical Handbook for Testators and Executors, including the new Death Duties. By C. E. STEWART, Esq., M.A., Barrister-at-Law. Price 1s. 6d.

Executors and Administrators, their Duties and Liabilities.
By G. F. EMERY, Barrister-at-Law.   Price 2s.

How to Appeal against your Rates
(In the Metropolis). By A. D. LAWRIE, Esq., M.A., Barrister-at-Law. Third Edition, revised and enlarged. Price 2s.

How to Appeal against your Rates
(Outside the Metropolis). By A. D. LAWRIE, Esq., M.A., Barrister-at-Law. Fifth and Enlarged Edition. Price 1s. 6d.

Investor's Book-keeping.
By EBENEZER CARR. Price 1s.

The Stockbroker's Handbook.
A Practical Manual for the Broker, his Clerk, and his Client. New Edition, with chapter on Options. Price 1s.

The Stockbroker's Correspondent.
Being a Letter-writer for Stock Exchange Business. Price 1s.

The Juryman's Handbook.
By SPENCER L. HOLLAND, Barrister-at-Law. Price 1s.

Income Tax; and how to get it Refunded.
Fourteenth and Revised Edition. By ALFRED CHAPMAN, Esq. Price 1s. 6d.

Land Tax: and how to get it Corrected.
With Appendix, containing Instructions to Assessors, 1897. By JOHN ARNOTT, F.S.I. Price 1s.

Inhabited House Duty: How and when to Appeal.
By ALFRED CHAPMAN, Esq. Price 1s.

Law of Water and Gas.
By C. E. STEWART, Esq., M.A., Barrister-at-Law. Price 1s. 6d.

The Law of Bankruptcy.
Showing the Proceedings from Bankruptcy to Discharge. By C. E. STEWART, Esq., Barrister-at-Law. Price 2s.

How to obtain a Divorce.
By NAPOLEON ARGLES, Esq., Solicitor. Price 1s. 6d.

Hoare's Mensuration for the Million;
Or, the Decimal System and its application to the Daily Employment of the Artizan and Mechanic. By CHARLES HOARE. Price 1s.

Ferguson's Buyers and Sellers' Guide; or, Profit on Return.
Showing at one view the Net Cost and Return Prices, with a Table of Discount. New and Rearranged Edition. Price 1s.

House-owners, Householders and Lodgers: their Rights and Liabilities as such. By J. A. DE MORGAN, Esq., Barrister-at-Law. Price 2s.

Bills of Sale.
By THOS. W HAYCRAFT, Esq., Barrister-at-Law. Price 2s. 6d.

The Wholesale and Retail Traders' Guide to the Law relating to the Sale and Purchase of Goods.
By C. E. STEWART, Esq., Barrister-at-Law. Price 1s. 6d.

Schonberg's Chain Rule:
A Manual of Brief Commercial Arithmetic. Price 1s.

County Council Guide. The Local Government Act, 1888.
By R. DENNY URLIN, Esq., Barrister-at-Law. Price 1s. 6d.

Lunacy Law.
An Explanatory Treatise on the Lunacy Act, 1890, for all who have the charge of, or are brought in contact with, persons of unsound mind. By D. CHAMIER, Esq., Barrister-at-Law. Price 1s. 6d.

Houses and Lands as Investments.
With Chapters on Mortgages, Leases, and Building Societies. By R. DENNY URLIN, Esq.. Barrister-at-Law. Price 1s.

How to Invest Money. Revised Edition. By E. R. GABBOTT. Price 1s.

From School to Office. Written for Boys. By F. B. CROUCH. Price 1s.

Pearce's Merchant's Clerk.
An Exposition of the Laws regulating the Operations of the Counting House. Twenty-first Edition. Price 2s.

The Theory of Book-keeping. By BENJAMIN SEEBOHM. Price 1s.

Double Entry; or, the Principles of Perfect Book-keeping.
By ERNEST HOLAH. Price 2s.

Powers, Duties and Liabilities of Directors under the Companies Acts 1862-1890.
By T. W. HAYCRAFT, Esq., Barrister-at-Law. Price 1s. 6d.

The Law of Innkeepers and the Licensing Acts.
By T. W. HAYCRAFT, Esq., Barrister-at-Law. Price 1s. 6d.

Validity of Contracts in Restraint of Trade.
By WILLIAM ARNOLD JOLLY, Barrister-at-Law. Price 1s.

Copyhold Enfranchisement with reference to the Copyhold Act, 1894. By ARTHUR DRAYCOTT. Price 1s.

Pawnbroker's Legal Handbook, based upon the Act of 1872.
By CHAN-TOON and JOHN BRUCE, Esqs., Barristers. Price 1s.

Criminal Evidence Act, 1898.
With Explanatory Notes. By CHARLES BRONTE MORGAN, Barrister-at-Law. 1s.

A Complete Summary of the Law Relating to the English Newspaper Press. Price 1s.

The Neutral Ship in War Time.
By A. SAUNDERS. Price 1s. net.

Schedule D of the Income Tax and how to Deal with it.
By S. W. FLINT. Price 1s. net.

Law Affecting the Turf, Betting and Gaming-Houses and the Stock Exchange.
By LAWRENCE DUCKWORTH, Barrister-at-Law. Price 1s.

Law Relating to Insurance Agents, Fire, Life, Accident and Marine.
By J. E. R. STEPHENS, Barrister-at-Law. Price 1s.

A HISTORY OF THE BANKING OF ALL NATIONS.

In Four Royal 8vo volumes, Price £5 net complete.
Now READY.

VOL. I. contains the History of Banking in the United States. By WILLIAM G. SUMNER.

VOL. II. contains History of Banking in Great Britain. By HENRY DUNNING MACLEOD. History of Banking in Russia. By ANT. E. HORN. History of Savings Banks in the United States. By JOHN P. TOWNSEND.

VOL. III. contains History of Banking in the Latin Nations, by PIERRE DES ESSARS; of the Banks of Alsace-Lorraine, by ARTHUR RAFFALOVICH; and in Canada, by BYRON E. WALKER.

VOL. IV. contains History of Banking in Germany and Austria-Hungary, Scandinavian Nations, Holland, China, Japan, etc.

ALDENHAM, LORD (H. H. GIBBS).
A Colloquy on Currency. New Edition (*in the press*).

ATTFIELD, J. B.
English and Foreign Banks : a Comparison.
Contents :—The Constitution of Banks; The Branch System ; The Functions of Banks. Price 3s. 6d. net.

AYER, JULES.
General and Comparative Tables of the World's Statistics. Area and Population, Religion, Finance, Currency, Army, Navy, Railways and Telegraphs, Capitals and Towns, Time at Capitals, etc., revised to end of March, 1899. On a sheet 35 × 22. Price 1s. net.

BARCLAY, ROBERT.
The Disturbance in the Standard of Value. Second and enlarged Edition. Price 2s.

BEAURE, Prof. A.
Manuel pratique de la Correspondance et des opérations de Commerce. (Part I.) Price 1s. 6d. net.

Partie appliquée, avec traité pratique des Opérations de Bourse. (Part II.) 3s. 3d. net.

Practical Mercantile Correspondence. A Collection of Business Letters. Price 2s. net.

Théorie et pratique de la Monnaie. Tome premier, Traité Théorique de la Monnaie et Statistique des Metaux Precieux. Price 3s. 6d. net.

Histoire de la Politique Monétaire statistique des Frappes et mouvement des Métaux précieux dans les principaux pays. Tome II. Price 5s. net.

BEEMAN, G. B., and FREDC. C. MATHIESON AND SONS.
Australian Mining Manual : a Handy Guide to the West Australian Market. Price 4s. net.

" Its shape and flexibility fit it for the side pocket, and the information it contains seems to be all that can be desired."—*Daily Chronicle*.

BIRKS, H. W.
Half-yearly Comparative Analysis of the Balance Sheets of London Joint Stock and Private Banks. Published February and August of each Year. Sheet Form, price 1s.; Book Form, bound leather, price 5s.

Investment Ledger. Designed for the Use of Investors. Bound leather. Price 3s. 6d.

BLACKWELL, P. T., B.A.
The Law relating to Factors : Mercantile Agents who sell and buy goods on commission, and who have goods entrusted to their care, including the Factors Act, 1889, and the repealed Factors Acts. Price 5s. net.

" It is a handy work, and brings the law on this subject within a moderate compass."—*Law Times*.

BOOTH, A. A., and M. A. GRAINGER.
Diagram for calculating the yield on Redeemable Stocks. Price 10s. 6d. net.

By means of a small ruler and a table of lines the true yield on a bond or stock purchased at a given price, which is redeemable either at or above par, can be obtained at once without calculation of any kind.

BORLASE, WILLIAM COPELAND, M.A.
Tin Mining in Spain, past and present. Price 2s. 6d.

BOSANQUET, BERNARD T.
Universal Simple Interest Tables, showing the Interest of any sum for any number of days at 100 different rates, from $\frac{1}{8}$ to $12\frac{1}{2}$ per cent. inclusive; also the Interest of any sum for one day at each of the above rates, by single pounds up to one hundred, by hundreds up to forty thousand, and thence by longer intervals up to fifty million pounds. 8vo, pp. 480. Price 21s. cloth.

BOURNE'S INSURANCE PUBLICATIONS.
Directory. Cloth gilt, price 5s.; post free, 5s. 6d. (annual).
Handy Assurance Manual. In Card cover, 1s., by post, 1s. 2d.; in Cloth cover, 1s. 6d., by post, 1s. 8d.; in Pocket-book, with convenient pocket, 2s. 6d., by post, 2s. 8d. (annual).
Guides. Published each month.

January—The Handy Assurance Guide—Seventeenth Year. February—Annual Bonuses. March—Expense Ratios of Life Offices. April—The Handy Assurance Guide. May—New Life Business and its Cost. June—The Handy Fire Insurance Guide. July—The Handy Assurance Guide. August—Valuation Summaries. September—Expense Ratios of Life Offices. October—The Handy Assurance Guide. November—New Life Business and its Cost. December—Premium Rates.

They are clearly printed on cards folding to 5 in. by 3 in., and giving in a singularly compact and convenient form the latest statistics of all the Offices. Price 3d., by post, $3\frac{1}{2}$d.; per dozen, 2s. 6d.; per 100, 16s. 8d

BROWNLEE'S
Handbook of British Columbia Mining Laws. For Miners and Prospectors. Price 1s.

BURGON, JOHN WILLIAM.
Life and Times of Sir T. Gresham. Including notices of many of his contemporaries. In two handsome large octavo volumes, embellished with a fine Portrait, and twenty-nine other Engravings. Published at 30s. Offered at the *reduced price of* 10s.

CARISS, ASTRUP.
Book-keeping by Double Entry: explaining the Science and Teaching the Art. Second Edition. Price 6s.

CASTELLI, C.
Theory of "Options" in Stocks and Shares. Price 2s. net.

CHAMIER, DANIEL.
Law relating to Literary Copyright and the Authorship and Publication of Books. Price 5s. net.

"The work may be conscientiously recommended for any one requiring a cheap and trustworthy guide."—*Athenæum*.

CHARLTON, R. H.
Useful Information for Gold Mining Investors. Price 1s.

CLARE, GEORGE.
A Money Market Primer and Key to the Exchanges. Second Edition, revised. Recommended by the Council of the Institute of Bankers. With Eighteen Full-page Diagrams. Price 5s.

COBB, ARTHUR STANLEY.
Threadneedle Street, a reply to "Lombard Street," and an alternative proposal to the One Pound Note Scheme sketched by Mr. Goschen at Leeds. Price 5s.

Mr. Goschen said at the London Chamber of Commerce, "Mr. Stanley Cobb proposes an alternative to my plan, and I recommended the choice between the two".

COUMBE, E. H., B.A. (Lond.).
A Manual of Commercial Correspondence. Including Hints on Composition, Explanations of Business Terms, and a large number of Specimen Letters as actually in current use, together with information on the General Commercial Subjects treated in the Correspondence. Price 2s. 6d. net.

CROSBIE, ANDREW, and WILLIAM C. LAW.
Tables for the Immediate Conversion of Products into Interest, at Twenty-nine Rates, viz.: From One to Eight per cent. inclusive, proceeding by Quarter Rates, each Rate occupying a single opening, Hundreds of Products being represented by Units. Second Edition, improved and enlarged. Price 12s. 6d.

CUMMINS, CHARLES.
$2\frac{3}{4}$ per cent. Interest Tables on £1 to £20,000 for 1 to 365 days. Price 5s. net.

Formation of the Accounts of Limited Liability Companies. Price 5s. net.

CUTHBERTSON, CLIVE, B.A.

A Sketch of the Currency Question. Price 2s. net.

"An admirable *resumé* of the controversy between monometallists and bimetallists."—*Times*.

DALE, BERNARD.

How are the Profits for the Year to be ascertained; or what is the Capital of a Company? Second and Revised Edition. Price 1s. net.

DEL MAR, ALEX.

History of the Monetary Systems in the various States. Price 15s. net.

> LIST OF CHAPTERS.—I. India from the Earliest Times. II. Ancient Persia. III. Hebrew Moneys. IV. Ancient Greece. V. Rome B.C. 369 to A.D. 1204. VI. The Sacred Character of Gold. VII. Pounds, Shillings and Pence. VIII. Gothic Moneys. IX. Moslem Moneys A.D. 622-1492. X. Early English Moneys. XI. Moneys of the Heptarchy. XII. Anglo-Norman Moneys. XIII. Early Plantagenet Moneys. XIV. Later Plantagenet Moneys. XV. The Coining Prerogative. XVI. Saxony and Scandinavia to Date. XVII. The Netherlands to Date. XVIII. Germany to Date. XIX. Argentine Confederation to Date. XX. Private Coinage.

The Science of Money. Second revised Edition. Demy 8vo, price 6s. net.

> CHAPTERS on—Exchange. Value as a Numerical Relation. Price. Money is a Mechanism. Constituents of a Monetary Mechanism. History of Monetary Mechanisms. The Law of Money. The Unit of Money is all Money. Money contrasted with other Measures. Limitation is the Essence of Moneys. Limitation: a Prerogative of the State. Universal Money a Chimera. Causes and Analysis of a Rate of Interest. Velocity of Circulation. Relation of Money to Prices. Increasing and Diminishing Moneys. Effects of Expansion and Contraction. The Precession of Prices. Revulsions of Prices. Regulation of Moneys.

DE SEGUNDO, E.

The Rapid Share Calculator. For Calculating $\frac{1}{8}$ths, $\frac{1}{16}$ths, and $\frac{1}{32}$nds. Price 10s. 6d. net.

"An ingenious mechanical contrivance for easily calculating fractional values."—*Standard*.

DONALD, T.

Accounts of Gold Mining and Exploration Companies. With Instructions and Forms for rendering the same to Head Office. Price 3s. 6d. net.

DRAPERS' ACCOUNTS.
A Manual for the Drapery and Allied Trades. By A CORPORATE ACCOUNTANT. Second Edition. Price 3s. 6d.

DUNCAN, W. W.
How Money makes Money. Price 2s. 6d. net.

DUNSFORD, F.
Handbook of Railway and other Securities for Fifteen Years, gives at a glance the Lowest and Highest Prices and Dividends paid. Published Annually. Price 1s. net.

EASTON, H. T.
Banks and Banking. Price 3s. 6d.
The Work of a Bank. Price 2s. net.

ELLIS, ARTHUR.
Rationale of Market Fluctuations. Third Edition. Price 7s. 6d.

ELLISON, THOMAS.
Cotton Trade of Great Britain. Including a History of the Liverpool Cotton Market and the Liverpool Cotton Brokers' Association. Price 15s.

EMERY, G. F., LL.M.
Handy Guide to Patent Law and Practice. Price 6s. net.

"The book is one which a layman will find extremely useful, and we can confidently recommend it also to solicitors."—*Law Notes.*

ENNIS, GEORGE, and ENNIS, GEORGE FRANCIS MACDANIEL.
The Registration of Transfers of Transferable Stocks, Shares, and Securities; with a Chapter on the Forged Transfers Act, and an Appendix of Forms. Price 7s. 6d.

"FENN ON THE FUNDS."
Being a Handbook of Public Debts. Containing Details and Histories of the Debts, Budgets and Foreign Trade of all Nations, together with Statistics elucidating the Financial and Economic Progress and Position of the various Countries. Sixteenth Edition, thoroughly Revised and in greater part Rewritten. Edited by S. F. VAN Oss, with the assistance of H. H. BASSETT. Demy 8vo, pp. 578. Price 15s.

"So much useful matter in any one volume is seldom to be met with."—*The Times.*

GABBOTT, E. R.
How to Invest in Mines: a Review of the Mine, the Company and the Market. Price 2s. 6d. net.

GARRATT, JOHN and CHARLES.
Exchange Tables, to convert the Moneys of Brazil, the River Plate Ports, Chili, Peru, Californian and Lisbon (Milreis and Reis, Dollars and Reals, Dollars and Cents) into British Currency, and *vice versâ*, varying by eighths of a penny. Price 10s. 6d.

GASKELL, W. H.
Silver Tables, showing relative equivalents of Bar Silver in London and New York. Vol. I.—From 47 cents to 67 cents; Vol. II.—From 67 cents to 87 cents, U.S. Currency; ascending by 1/8th, at Exchange of $4.80 to $4.90 per £ sterling, ascending by 1/4th of a cent. Price 15s., 2 vols.; or if sold separately, price 10s. each.

GEORGE, E. MONSON.
Railways in India; their Economical Construction and Working. Price 2s. 6d.

GIBBS, Hon. HERBERT.
A Bimetallic Primer. Third Edition, revised. Price 1s. net.

GILBERT.
Interest and Contango Tables. Price 10s net

GOLDMANN, CHARLES SYDNEY, F.R.G.S., with the co-operation of JOSEPH KITCHIN.
South African Mines: giving the Position, Results and Developments of all South African Mines; together with an Account of Diamond, Land, Finance and kindred concerns. In three volumes.

VOL. I.—Devoted to detailed descriptions of all Witwatersrand Mining Companies, containing about 500 pages.

VOL. II.—Dealing with Mining Companies other than Rand, together with Rhodesian, Diamond, Finance, Investment, Land, and Miscellaneous Companies. It contains about 220 pages.

VOL. III.—100 Maps and Plans of Mining Properties, including a large Scale Map of the Rand in seventeen sections, together with dip, tonnage and other charts.

Price (net) £3 3s.

GOLDMANN, CHARLES SYDNEY.
The Financial, Statistical and General History of the Gold and other Companies of Witwatersrand, South Africa. Price 12s. 6d. net.

GOODRICKE, C. A., & CO.
Tea Exchange Tables. Showing the conversion of London gross sale price into Calcutta gross sale price and *vice versa*. Price 21s. net.

GOSCHEN, the Right Hon. GEO. J., M.P.
Theory of Foreign Exchanges. Sixteenth Edition. 8vo. Price 6s.

GREVILLE, M. E.
From Veld and " Street ". Rhymes more or less South African. Price 1s.

GUMERSALL.
Tables of Interest, etc. Interest and Discount Tables, computed at 2½, 3, 3½, 4, 4½ and 5 per cent., from 1 to 365 days and from £1 to £20,000; so that the Interest or Discount on any sum, for any number of days, at any of the above rates, may be obtained by the inspection of one page only.
Nineteenth Edition, in 1 vol., 8vo (pp. 500), price 10s. 6d., cloth, or strongly bound in calf, with the Rates per Cent. cut in at the fore-edge, price 16s. 6d.

HAM'S
Customs Year-Book. A new List of Imports and Exports, with Appendix and a brief account of the Ports and Harbours of the United Kingdom. Published Annually. Price 3s.; with Warehousing Supplement, 4s. 6d. net.
Inland Revenue Year-Book. The recognised book of Legal Reference for the Revenue Departments. Published Annually. Price 3s.; with Warehousing Supplement, 4s. 6d. net.

HAM, PANTON.
Universal Interest Table. For calculating Interest at any Rate on the Moneys of all Countries. Price 2s. 6d. net.

HARLOW.
Examination Questions in Book-keeping. Price 2s. 6d.

HAUPT, OTTOMAR.
Arbitrages et Parités. Traité des Opérations de Banque, contenant les usages commerciaux, la théorie des changes et monnaies, et la statistique monétaire de tous les pays du globe. Huitième édition. Price 12s. 6d. net.
The Monetary Question in 1892. Price 5s.

HAWKE, H. E.
Instantaneous Share Reckoner. Price 2s.

HENSELIN, ADOLF.
Ready Reckoner, by which multiplication of factors from 1 × 1 to 999 × 999 can be seen at a glance, and those of still larger numbers can be effected with the utmost rapidity. By these tables the division of any one number by another can also be done. Together with Calculating Tables for circles. Price 8s. net.

HIBBERT, W. NEMBHARD, LL.D.
Law relating to Company Promoters. Price 5s. net.

HIGGINS, LEONARD R.
The Put-and-Call. Price 3s. 6d. net.

HOWARTH, WM.
Our Banking Clearing System and Clearing Houses. Third and Enlarged Edition. Price 3s. 6d.

HUGHES, T. M. P.
Investors' Tables for ascertaining the true return of Interest on Investments in either Permanent or Redeemable Stocks or Bonds, at any rate per cent., and Prices from 75 to 140. Price 6s. 6d. net.

HUNTER'S
Motor, Cycle, and Component Parts Official Intelligence. Price 5s.

HUTCHISON, JOHN.
Practice of Banking; embracing the Cases at Law and in Equity bearing upon all Branches of the Subject. Volumes II. and III. Price 21s. each. Vol. IV. Price 15s.

INGRAM.
Improved Calculator, showing instantly the Value of any Quantity from One-sixteenth of a Yard or Pound to Five Hundred Yards or Pounds, at from One Farthing to Twenty Shillings per Yard or Pound. Price 7s. 6d.

JACKSON, GEORGE.
Book-keeping. A Check-Journal; combining the advantages of the Day-Book, Journal and Cash-Book; forming a complete System of Book-keeping by Double Entry; with copious illustrations of Interest Accounts and Joint Adventures; and a method of Book-keeping, or Double Entry by single.

Twenty-first Edition, with the most effectual means of preventing Fraud, Error and Embezzlement in Cash Transactions, and in the Receipt and Delivery of Goods, etc. Price 5s.

JOHNSON, GEORGE, F.S.S., A.I.S.
Mercantile Practice. Deals with Account Sales, Shipping, Exchanges, Notes on Auditing and Book-keeping. Price 2s. 6d. net.

JONES, CHARLES.
The Solicitor's Clerk: the Ordinary Practical Work of a Solicitor's Office. Fifth and Revised Edition. Price 2s. 6d. net.

Companion to the Solicitor's Clerk. A continuation of the "Solicitor's Clerk," embracing Magisterial and Criminal Law, Licensing, Bankruptcy Accounts, Bookkeeping, Trust Accounts, etc. (*Second and Revised Edition.*) Price 2s. 6d. net.

The Business Man's County Court Guide. A Practical Manual, especially with reference to the recovery of Trade Debts. Second and Revised Edition. Price 2s. 6d. net.

Book of Practical Forms for Use in Solicitors' Offices. Containing over 400 Forms and Precedents in the Queen's Bench Division and the County Court. *Price* 5s. *net.*

JONES, HUGH.
A Guide to the Liquor Licensing Acts. Price 2s. 6d. net.

JOURNAL OF THE INSTITUTE OF BANKERS.
Monthly, 1s. 6d.

JUDICIAL TRUSTEES ACT, 1896.
And the Rules made thereunder. By A. SOLICITOR. Price 2s. 6d. net.

KELLY and WALSH.
Dollar or Taels and Sterling Exchange Tables. Compiled to facilitate Exchange Calculations at the finer rates at which Eastern business is now done. At different rates from 1s. 6d. to 3s. 4d., advancing by Sixteenths of a Penny. Price 10s. 6d. net.

KERR, ANDREW WILLIAM, F.S.A. (Scot.).
Scottish Banking during the Period of Published Accounts, 1865-1896. Price 5s.

KINDELL'S
African Market Manual, giving particulars of all Companies (African or Rhodesian) dealt in the African Market of the Stock Exchange. April, 1899. Price 5s. net.

KÖLKENBECK, ALFRED.
Rates of Stamp Duties on Bills of Exchange all over the World. Price 1s. net.

LECOFFRE, A.
Tables of Exchange between France, Belgium, Switzer land and Great Britain; being French Money reduced inte English from 25 francs to 26 francs per pound sterling in Rates each advancing by a quarter of a centime, showing the value from one franc to one million of francs in Englisl Money. 21s.

Tables of Exchange between Austria, Holland and Great Britain. Price 15s.

LEWIS, WILLIAM.
Tables for finding the Number of Days, from one day to any other day in the same or the following year. Price 12s. 6d.

LITHIBY, JOHN.
The Law of District and Parish Councils. Being the Local Government Act, 1894, with an Appendix containing Numerous Statutes referred to in, or incorporated with the Act itself; including the Agricultural Gangs Act, the Agricultural Holdings Act, the Allotments Acts, Baths and Washhouses Acts, Burial Acts, Fairs Acts, Infant Life Protection Act, Knackers Acts, Lighting and Watching Act, Public Improvements Act, Public Libraries Acts, and numerous Extracts from the Public Health Act, 1875, and other Statutes. Also the Orders and Circulars of the Local Government Board, together with copious Notes and a ful Index. Second Edition, revised and enlarged. Demy 8vc 659 pages. Price 15s.

LYNCH, H. F.
Redress by Arbitration; being a Digest of the Lav relating to Arbitration and Award. Third and Revise Edition. Price 5s.

MACFEE, K. N., M.A.
Imperial Customs Union. A practical Scheme of Fisca Union for the purposes of Defence and Preferential Trade from a Colonist's Standpoint. Price, cloth, 2s. 6d.; paper 1s. 6d.

McEWEN'S
Bankruptcy Accounts. How to prepare a Statemen of Affairs in Bankruptcy. A Guide to Solicitors and others Price 2s. 6d.

MARRACK, RICHARD, M.A.
The Statutory Trust Investment Guide. The par ticulars as to Investments eligible, compiled and arrange by Fredc. C. Mathieson and Sons. Second Edition revised and enlarged. Price 6s. net.

"We think the authors have executed their task well, and that their book will b found useful. We have often thought that a lawyer and a practical man writing i concert might produce a very excellent book."—*Law Quarterly Review.*

MATHIESON, FREDC. C., & SONS.

Monthly Traffic Tables; showing Traffic to date and giving as comparison, the adjusted Traffics of the corresponding date in the previous year. Price 6d., by post 7d. Monthly.

American Traffic Tables. Monthly. Price 6d., by post, 7d.

Highest and Lowest Prices, and Dividends paid during the past six years. Annually. Price 2s. 6d.

Provincial Highest and Lowest Prices as quoted on the following Stock Exchanges: Birmingham, Dublin, Edinburgh, Glasgow, Leeds, Liverpool, Manchester and Sheffield. Annually. Price 2s. 6d.

Six Months' Prices and Dates. Uniform with "Highest and Lowest Prices". Annually, in July. Price 2s. 6d.

Investor's Handbook of Railway Statistics, 1878-1899. Annually. Price 6d.

Investor's Ledger. Price 3s. 6d.

Indian Railway Companies. A Handbook for Officials, Stockbrokers and Investors. Annually. Price 1s.

Monthly Mining Handbook. Price 1s.

Redeemable Investment Tables. Calculations checked and extended. By A. SKENE SMITH. Price 15s. net.

MAUDE, WILLIAM C., Barrister-at-Law.

Property Law for General Readers. Price 3s. 6d.

MAY, J. R.

Institute of Bankers' Examinations. Examination Questions in Arithmetic and Algebra. Preliminary and Final for Nineteen Years, 1880-1898, with Answers. Preliminary Part, price 1s. 6d. Final Part, price 1s. 6d.

MELIOT'S

English and French Explanatory Dictionary of Terms and Phrases relating to Finance, the Stock Exchange, Joint-Stock Companies and Gold Mining. Price 5s. net.

MELSHEIMER and GARDNER.

Law and Customs of the London Stock Exchange. Third Edition. Price 7s. 6d.

MERCES, F. A. D.
Indian Exchange Tables. A New Edition, Showing the Conversion of English Money into Indian Currency, and *vice versa*, calculated for every Thirty-second of a Penny; from 1s. to 1s. 6d., price 15s. net.; Supplements 1/ to 1/ 31/32 5s.; 1/1 to 1/1 31/32 5s. net.

Indian Interest Tables, from 1 to 15 per cent. per annum of 360 and 365 days; also Commission, Discount and Brokerage from 1 anna to 15 per cent. Price 8s. net.

Indian Ready Reckoner. Containing Tables of Rates by Number, Quantity, Weight, etc., including fractions of a Maund, at any rate from $\frac{1}{2}$ Pie to 250 Rs.; also Tables of Income, Exchange (1s. 2d. to 1s. 8d.), Interest and Commission. Sixth Edition. Price 36s. net.

MILFORD PHILIP.
Pocket Dictionary of Mining Terms. Third Edition. Price 1s.

MINTON-SENHOUSE, R. M., and EMERY, G. F.
The Workmen's Compensation Act. Price 15s.

NORMAN, F. S. C.
Tables of Commission and Due Dates. Price 2s. net.

NORMAN, J. H.
Universal Cambist. A Ready Reckoner of the World's Foreign and Colonial Exchanges of Seven Monetary and Currency Intermediaries, also the Present Mechanism of the Interchanges of Things between Man and Man and between Community and Community. Price 12s. 6d. net.

PALGRAVE, R. H. INGLIS.
Bank Rate in England, France and Germany, 1844-1878, with Remarks on the Causes which influence the Rate of Interest charged, and an Analysis of the Accounts of the Bank of England. Price 10s. 6d.

PAULL, J.
Columbia and Klondyke Market Manual. Price 2s. 6d.

PETHERICK, EDWARD A.
Australia in 1897. The Country and its Resources, Population, Public Works and Finances. With Two Maps. Second Edition. Paper, 1s. 6d.; cloth, 2s. 6d.

PHILLIPS, MABERLY.
A History of Banks, Bankers and Banking in Northumberland, Durham and North Yorkshire, illustrating the commercial development of the North of England from 1755 to 1894. With numerous Portraits, Fac-similes of Notes, Signatures, Documents, etc. Price 31s. 6d.

POCOCK, W. A.
An Epitome of the Practice of the Chancery and Queen's Bench Divisions of the High Court of Justice. Price 2s. 6d. net.

POOR, H. V. & H. W.
Manual of the Railroads of the United States, and other Investment Securities.
Statements showing the Financial Condition, etc., of the United States, and of all leading Industrial Enterprises.
Statements showing the Mileage, Stocks, Bonds, Cost, Traffic, Earnings, Expenses and Organizations of the Railroads of the United States, with a Sketch of their Rise Progress, Influence, etc. Together with 70 Maps and an Appendix, containing a full Analysis of the Debts of the United States and of the several States, published Annually. Price 42s.
The Money Question. A Handbook for the Times. Price 6s. net.

PROBYN, L. C.
Indian Coinage and Currency. Price 4s.

QUESTIONS ON BANKING PRACTICE.
Revised by, and issued under the sanction of, the Council of the Institute of Bankers. Fifth Edition, revised and enlarged. Price 6s.

RAIKES, F. W. (His Honour JUDGE), Q.C., LL.D.
The Maritime Codes of Holland and Belgium. Price 10s. 6d.
The Maritime Codes of Spain and Portugal. Price 7s. 6d. net.

"Dr. Raikes is known as a profound student of maritime jurisprudence, and he has been able to use his knowledge in a number of notes, in which the law of England and of other countries is compared with that of the Iberian Peninsula."—*Law Journal*.

RICHARDSON, G. H.
Book-keeping for Weekly Newspapers. A Manual for Newspaper Managers and Clerks. With an Introduction by H. CALDER MARSHALL, Chartered Accountant. Second Edition, entirely revised and enlarged. Price 3s. 6d.

RICHTER, HENRY.
The Corn Trade Invoice Clerk. Price 1s. net.

ROBINSON.
Share and Stock Tables; comprising a set of Tables for Calculating the Cost of any number of Shares, at any price from 1-16th of a pound sterling, or 1s. 3d. per share, to £310 per share in value; and from 1 to 500 shares, or from £100 to £50,000 stock. Seventh Edition, price 5s.

ROYLE, WILLIAM.
Laws relating to English and Foreign Funds, Shares and Securities. The Stock Exchange, its Usages, and the Rights of Vendors and Purchasers. Price 6s.

RUTTER, HENRY.
General Interest Tables for Dollars, Francs, Milreis, etc., adapted to both the English and Indian Currency, at rates varying from 1 to 12 per cent, on the Decimal System. Price 10s. 6d.

SAWYER, JOHN.
Practical Book-keeping. Suitable for all Businesses. Price 2s. 6d.

SCHULTZ.
Universal American Dollar Exchange Tables, Epitome of Rates from $4.80 to $4.90 per £, and from 3s. 10d. to 4s. 6d. per $, with an Introductory Chapter on the Coinages and Exchanges of the World. Price 10s. 6d.

Universal Dollar Tables. Complete United States Edition. Covering all Exchanges between the United States and Great Britain, France, Belgium, Switzerland, Italy, Spain and Germany. Price 21s.

Universal Interest and General Percentage Tables on any given amount in any Currency. Price 7s. 6d.

English-German Exchange Tables, from 20 marks to 21 per £ by ·025 mark per £, progressively. Price 5s.

SHEARMAN, MONTAGUE, and THOS. W. HAYCRAFT.
London Chamber of Arbitration. A Guide to the Law and Practice, with Rules and Forms. Second Edition. Price 2s. 6d.

SHEFFIELD, GEORGE.
Simplex System of Solicitors' Book-keeping. Price 3s. 6d. net.

SIMONSON, PAUL F., M.A. (Oxon.).
Treatise on the Law Relating to Debentures and Debenture Stock issued by Trading and Public Companies and by Local Authorities, with Forms and Precedents. Second and Revised Edition. Price 21s.

SMITH, A. SKENE.
Compound Interest: as exemplified in the Calculation of Annuities, immediate and deferred, Present Values and Amounts, Insurance Premiums, Repayment of Loans, Capitalisation of Rentals and Incomes, etc. Second and Enlarged Edition. Price 1s. net.

"It is written with a business-like explicitness, and cannot fail to prove useful."—*Scotsman.*

SMITH, JAMES WALTER.
The Law of Banker and Customer. New and Revised Edition. Price 5s.

STEPHENS, T. A.
A Contribution to the Bibliography of the Bank of England. Price 10s. 6d.

STEVENS, W. J.
Home Railways as Investments. Second Edition, 1897. Price 2s. 6d. net.
"An interesting and instructive treatise."—*Daily Chronicle.*

STEWART, F. S.
English Weights, with their equivalents in Kilogrammes. Calculated from 1 pound to 1 ton by pounds, and from 1 ton to 100 tons by tons. Compiled expressly for the use of Merchants and Shipping Agents to facilitate the making out the Documents for Foreign Custom Houses. Price 2s. 6d. net.

STOCK EXCHANGE OFFICIAL INTELLIGENCE;
Being a carefully compiled *précis* of information regarding British, American and Foreign Stocks, Corporation, Colonial and Government Securities, Railways, Banks, Canals, Docks, Gas, Insurance, Land, Mines, Shipping, Telegraphs, Tramways, Water-works and other Companies. Published Annually under the sanction of the Committee. Price 50s.

STRONG, W. R.
Short-Term Table for apportioning Interest, Annuities, Premiums, etc., etc. Price 1s.

STUTFIELD, G. HERBERT, and CAUTLEY, HENRY STROTHER.
Rules and Usages of the Stock Exchange. Containing the Text of the Rules and an Explanation of the general course of business, with Practical Notes and Comments. Second and Revised Edition. Price 5s.

TATE.
Modern Cambist. A Manual of Foreign Exchanges. The Modern Cambist: forming a Manual of Foreign Exchanges in the various operations of Bills of Exchange and Bullion, according to the practice of all Trading Nations; with Tables of Foreign Weights and Measures, and their Equivalents in English and French.
"A work of great excellence. The care which has rendered this a standard work is still exercised, to cause it to keep pace, from time to time, with the changes in the monetary system of foreign nations."—*The Times.*
Twenty-third Edition. By HERMANN SCHMIDT. Price 12s.
Counting House Guide to the Higher Branches of Commercial Calculation. Price 7s. 6d.

TAYLER, J.
Red Palmer. A Practical Treatise on Fly Fishing. Third Edition. Price 1s. 6d.

TAYLER, J.—*(continued).*
A Guide to the Business of Public Meetings. The Duties and Powers of Chairman, with the modes of Procedure and Rules of Debate. Second Edition. Price 2s. 6d. net.
The Public Man: His Duties, Powers and Privileges, and how to Exercise them. Price 3s. 6d. net.

VAN DE LINDE, GERARD.
Book-keeping and other Papers. Adopted by the Institute of Bankers as a Text-Book for use in connection with their Examinations. New and Enlarged Edition. Price 6s. 6d. net.

VAN OSS, S. F.
American Railroads and British Investors. Price 3s. 6d. net.
Stock Exchange Values: A Decade of Finance, 1885-1895. Containing Original Chapters with Diagrams and Tables giving Reviews of each of the last Ten Years—Trade Cycles—The Course of Trade, 1884 to 1894—Silver—New Capital Created, 1884 to 1894—The Money Market, 1884 to 1894—Government and Municipal Securities—Colonial Securities—Foreign Government Securities—Home Railway Stocks—American Railways—Foreign and Colonial Railways and Miscellaneous Securities. Together with Charts showing at a glance prices of principal securities for past ten years, and Highest and Lowest Prices year by year (1885 to 1894 inclusive) of every security officially quoted on the Stock Exchange, with dates and extreme fluctuations (extending to over 200 pages of Tables), compiled by Fredc. C. Mathieson & Sons. Price 15s. net.

"An unusually interesting chronicle of financial events during the last ten years. ... We have not anywhere come across one so concise and yet so complete."—*Athenæum.*

WARNER, ROBERT.
Stock Exchange Book-keeping. Price 2s. 6d. net.

WILEMAN, J. P., C.E.
Brazilian Exchange, the Study of an Inconvertible Currency. Price 5s. net.

WILHELM, JOHN.
Comprehensive Tables of Compound Interest (not Decimals) on £1, £5, £25, £50, £75 and £100. Showing Accumalations Year by Year for Fifty Years at Rates of Interest from 1 (progressing $\frac{1}{4}$) to 5 per cent. Price 2s. 6d. net.

WILLDEY.
Parities of American Stocks in London, New York and Amsterdam, at all Rates of Exchange of the day. Price 2s.

WILSON.
Author's Guide. A Guide to Authors; showing how to correct the press, according to the mode adopted and understood by Printers. On Card. Price 6*d*.

Investment Table: showing the Actual Interest or Profit per cent. per annum derived from any purchase or investment at rates of Interest from 2½ to 10 per cent. Price 2*s*. net.

WOODLOCK, THOMAS F.
The Anatomy of a Railroad Report. Price 2*s*. 6*d*. net.
"Careful perusal of this useful work will enable the points in an American railroad report to be grasped without difficulty."—*Statist*.

RECENT PAMPHLETS.

Indian Currency: An Essay.
By WILLIAM FOWLER, LL.B. Price 1*s*.

Notes on Money and International Exchanges.
By SIR J. B. PHEAR. Price 1*s*.

Company Law.
A Short Comparison of the Two Bills now before Parliament. By GEO. ST. JOHN MILDMAY, Barrister-at-Law. Price 6*d*.

The Indian Finance Difficulty.
A Solution. Price 6*d*.

Suggested Alterations in the Bank Act of 1844.
By an EX-BANK MANAGER. Revised Edition. Price 1*s*.

Cost Price Life Assurance.
A Plain Guide to Offices yielding 2 and 4 per cent. Compound Interest per annum on Ordinary and Endowment Policies. By T. G. ROSE. Price 6*d*.

The Currency of China.
(A Short Enquiry). By JAMES K. MORRISON. Price 1*s*.

Pamphlets, etc., on Bimetallism.

BULL'S CURRENCY PROBLEM AND ITS SOLUTION. Cloth, 2*s*. 6*d*.

DICK'S INTERNATIONAL BULLION MONEY. Price 6*d*.

DOUGLAS (J. M.) GOLD AND SILVER MONEY: A Vital British Home Question, with Tables of Average Prices of Commodities and Silver from 1846 till 1892. Price 6*d*.

ELLISEN'S ERRORS AND FALLACIES OF BIMETALLISM. Price 6*d*.

MONEY, WHAT IS IT? AND WHAT IS ITS USE? Price 1*s*.

GEORGE'S THE SILVER AND INDIAN CURRENCY QUESTIONS. Price 1s. 3d.

LEAVER'S MONEY: its Origin, its Internal and International Rise and Development. Price 1s.

MANISTY'S CURRENCY FOR THE CROWD; or, Great Britain Herself Again. Price 1s.

MEYSEY-THOMPSON'S (SIR HENRY M., Bart., M.P.) PRIZE ESSAY. Injury to British Trade and Manufactures. By GEO. JAMIESON, Esq. Price 6d.

MILLER'S DISTRIBUTION OF WEALTH BY MONEY. Price 1s.

MONOMETALLISM UNMASKED; OR THE GOLD MANIA OF THE NINETEENTH CENTURY. By A SENIOR OPTIME. 6d.

NORMAN'S PRICES AND MONETARY AND CURRENCY EXCHANGES OF THE WORLD. Price 6d.

NORMAN'S SCIENCE OF MONEY. Price 1s.

NORMAN'S THE WORLD'S TWO METALS AND FOUR OTHER CURRENCY INTERMEDIARIES. Price 1s.

SCHMIDT'S SILVER QUESTION IN ITS SOCIAL ASPECT. An Enquiry into the Existing Depression of Trade and the present position of the Bimetallic Controversy. By HERMANN SCHMIDT. Price 3s.

SCHMIDT'S INDIAN CURRENCY DANGER. A criticism of the proposed alterations in the Indian Standard. Price 1s. 6d.

SEYD'S SILVER QUESTION IN 1893. A Simple Explanation. By ERNEST SEYD, F.S.S. Price 2s., cloth.

SEYD'S BIMETALLISM IN 1886; AND THE FURTHER FALL IN SILVER. By ERNEST J. F. SEYD. Price 1s.

SMITH'S BIMETALLIC QUESTION. By SAM. SMITH, Esq., M.P. Price 2s. 6d.

SOWERBY'S THE INDIAN RUPEE QUESTION AND HOW TO SOLVE IT. Price 6d.

THE GOLD STANDARD. A Selection of Papers issued by the Gold Standard Defence Association in 1895-1898 in Opposition to Bimetallism. Price 2s. 6d.

THE GOLD BUG AND THE WORKING MAN. Price 6d.

TWIGG'S PLAIN STATEMENT OF THE CURRENCY QUESTION, with Reasons why we should restore the Old English Law of Bimetallism. Price 6d.

ZORN'S THEORY OF BIMETALLISM. Price 3d.

AGER'S TELEGRAM CODES.

THE A Y Z TELEGRAM CODE.

Consisting of nearly 30,000 Sentences and Prices, etc., with a liberal supply of spare words, for the use of Bankers, Brokers, Manufacturers, Merchants, Shippers. etc. The Code words carefully compiled from the "Official Vocabulary". The Active Stocks quoted on the Stock Exchange, London, with a list of American Bonds having Code words to them, makes this a useful Code for Stockbrokers. Price 16s. net.

"It forms a handy volume, compiled with evident care and judgment, and clearly and correctly printed."—*Daily Chronicle.*

"All the sentences in each par. are alphabetically arranged, so that it should not be difficult to code a telegram expeditiously and to interpret a code message upon receipt should even be easier."—*Daily Telegraph.*

THE SIMPLEX STANDARD TELEGRAM CODE.

Consisting of 205,500 Code Words. Carefully compiled in accordance with latest Convention rules. Arranged in completed hundreds. Printed on hand-made paper; strongly bound. Price £5 5s.

THE DUPLEX COMBINATION STANDARD CODE.

Consisting of 150,000 Words.

With a Double Set of Figures for every Word, thus affording opportunity for each Figure System of Telegraphing to be used. Every word has been compiled to avoid both literal and telegraphic similarities. Price £4 4s.

The Extension Duplex Code of about 45,000 more Words.

These are published with the view to being either used in connection with the "Duplex," or for special arrangement with the Figure System for PRIVATE CODES by agreement. Price £1 1s.

THE COMPLETE DUPLEX CODE,

Of 195,000 Words in Alphabetical and Double Numerical Order, *i.e.*, the above two Codes bound together. Price £5 5s.

Ager's Standard Telegram Code of 100,000 Words.

Compiled from the Languages sanctioned at the Berlin Telegraph Convention. Price £3 3s.

Ager's Standard Supplementary Code for General Merchants.
The 10,250 Words with sentences. In connection with the "Standard". Price 21s.

Ager's Telegram Code. 56,000 good Telegraphic Words, 45,000 of which do not exceed eight letters. Compiled from the languages sanctioned by the Telegraph Convention. Third Edition. Price £2 2s.

Ager's Alphabetical Telegram Code. The Code Words in sequence to the 150,000 Words in the Duplex Standard Code. Price 25s. Two or more copies, 21s. each.
N.B.—Can also be obtained bound up with the Duplex or Prefix Code.

Ager's Telegraphic Primer. With Appendix. Consisting of about 19,000 good English and 12,000 good Dutch Telegraphic Words. 12,000 of these have sentences. Price 12s. 6d.

Ager's General and Social Code, For Travellers, Brokers, Bankers and Mercantile Agents. Price 10s. 6d.

TELEGRAPH CODES.

Anglo-American Cable Code. Price 21s.

Broomhall's Comprehensive Cipher Code.
Mining, Banking, Arbitrage, Mercantile, etc. Arranged for nearly 170,000 Phrases. Price £3 13s. 6d., cloth. Limp leather, price £4 4s.

Clauson-Thue's A B C Universal Commercial Electric Telegraphic Code,
Adapted for the Use of Financiers, Merchants, Shipowners, Brokers, Agents, etc. Fourth Edition. Price 15s. net.

Clauson-Thue's A 1 Universal Electric Telegraph Code,
For the Use of Financiers, Merchants, Shipowners, Underwriters, Engineers, Brokers, Agents, etc. Price 25s. net.

Figure Code for Stocks and Shares.
To be used with the "Official Vocabulary," or any similar list of numbered Words. Price 42s.

Hawke's Premier Cyphêr Telegraphic Code.
Price 10s. 6d. See back page of this Catalogue.

McNeill's Mining and General Telegraph Code.
Arranged to meet the requirements of Mining, Metallurgical and Civil Engineers, Directors of Mining and Smelting Companies, Bankers, Brokers, Solicitors and others. Price 21s. net.

Moreing and McCutcheon's General Commercial and Mining Telegram Code.
Comprising 274,000 Words and Phrases. Price £5 5s. net.

Moreing and Neal's General and Mining Code.
For the Use of Mining Companies, Mining Engineers, Stockbrokers, Financial Agents, and Trust and Finance Companies. Price 21s.

Official Vocabulary in Terminational Order.
Price 40s. net.

One-word "Firm Offer" Telegraphic Code with One-word "5 Offers" Reply Code. Price 7s. 6d.

Scott's Shipowners' Telegraphic Code.
New Edition. 1896. Price 21s.

Stockbrokers' Telegraph Code. Price 5s. net.

Watkins' Ship-broker's Telegraph Code.
Price £4 net. Two copies, £7 net.

Whitelaw's Telegraph Cyphers. 338,200 in all.
400,000 Cyphers in one continuous alphabetical order. Price £12 10s.

202,600 words, French, Spanish, Portuguese, Italian
    and Latin. Price . . . . 150s. each net.
53,000 English words . . . . . 50s. ,, ,,
42,600 German ,, ... 50s. ,, ,,
40,000 Dutch ,, ... 50s. ,, ,,

338,200
68,400 Latin, etc., etc. (Original Edition), included in the above 202,600 . . . 60s. ,, ,,
25,000 English (Original Edition), included in the above 53,000 . . . . . 40s. ,, ,,
22,500 of the English words arranged 25 to the page, with the full width of the quarto page for filling in phrases . . 60s. ,, ,,
14,400 of the Latin words arranged so as to represent any 3-letter group, or any three 2-figure groups up to 24 . . 15s. ,, ,,

Willink's Public Companies' Telegraph Code.
Price 12s. 6d. net.

www.ingramcontent.com/pod-product-compliance
Lightning Source LLC
Chambersburg PA
CBHW031812230426
43669CB00009B/1105